Intellectual Property *for* Integrated Circuits

Kiat Seng Yeo § Kim Tean Ng
Zhi Hui Kong § Tricia Bee Yoke Dang

ISBN 978-1-932159-85-1

Printed and bound in the U.S.A. Printed on acid-free paper
10 9 8 7 6 5 4 3 2 1

Library of Congress Cataloging-in-Publication Data

Library of Congress Cataloging-in-Publication Data for *Intellectual Property for Integrated Circuits* can be found on the J. Ross Publishing website at www.jrosspub.com/wav in the WAV download section for this book.

Direct all inquiries to J. Ross Publishing, Inc., 5765 N. Andrews Way, Fort Lauderdale, FL 33309.

Phone: (954) 727-9333
Fax: (561) 892-0700
Web: www.jrosspub.com

Contents

Acknowledgments

No one walks alone. In fact, in every important passage of our lives, there are always bound to be some special individuals whose constant support, assistance, encouragement, or mere companionship along the way, are crucial enough to leave an indelible mark on our route to success. The timely and smooth production of this book is no different in that respect. We cherish them all.

Kiat Seng YEO is grateful and thankful to his wife Ah Cheng and their three children Shun Yuan, Shun Yi, and Shun Yu. Their continuous support, patience, and encouragement were instrumental in the successful completion of this book.

Kim Tean NG in writing the relevant chapters of this book benefited from the generous assistance of his fellow colleagues in Nanyang Law LLC, to whom he is grateful and deeply indebted. He is also thankful to his wife Kim Bee for her unwavering support and patience; he dedicates this work to her.

Throughout the process of writing this manuscript, Zhi Hui KONG was truly blessed to have received abundant help and support from a number of important individuals. She would like to convey her heartfelt appreciation to her parents and loved ones for steadfastly standing by her throughout the process of writing this book. Words can never express the warmth and thankfulness she feels for each one of them. Her deepest gratitude goes to her loving husband, Kam Chew, for his enormous support and endless love, patience, and encouragement. Not only is he the source of inspiration that has geared her forward while researching this project, he is also her unwavering pillar of support both emotionally and intellectually.

Tricia Bee Yoke DANG is thankful to her family for their support, encouragement, and inspiration throughout this project. She is also thankful to Tim Pletscher at J. Ross Publishing for his patience and understanding in waiting for the delivery of the manuscript.

We are truly thankful to Tim Pletscher, Senior Acquisitions Editor of J. Ross Publishing, and his team for the professional production of this book. We wish to thank them for their dedicated and tireless efforts in bringing this book into reality.

We wish to express our appreciation to Nanyang Technological University and Nanyang Law LLC for creating an environment conducive to education, research, intrinsic motivation, collaboration, and innovation.

Finally, the success of this book would not have been possible without the kind assistance of our researchers and students. In particular, we would like to thank Lih Chieh PNG, Wee Khee HO, and Lin Biao WANG for their enthusiastic involvement and contribution in the preparation of the manuscript.

Preface

Thank you for picking up this book.

Now you might start to wonder: Is this book for me? Let us help you to decide. Please read on.

Are you a scientist, researcher, or an engineer seeking to develop your knowledge of Intellectual Property (IP) for Integrated Circuits (ICs)? Are you an IP practitioner or attorney keen to specialize in the ever-growing field of ICs? Are you an undergraduate or postgraduate yearning to enrich your learning on IP for ICs even if you have zero knowledge on this at this very moment? Have you been searching high and low for an IP book that is customized for the IC industry?

If your answer to any of the questions above is a "yes," end your search and look no further. This is the book for you.

Compelling Features of the Book

Entitled *Intellectual Property for Integrated Circuits,* this book has been conceived as an effort to merge IP and ICs. The field of IP is rapidly gaining recognition. Meanwhile, the worldwide IC industry's growth potential is tremendous. The knowledge-intensive IC industry is an area that incessantly provides a realm for innovation, and therefore, its chemistry with IP law is important. Ironically, despite the accelerating importance of IP rights and the IC industry, the book that specifically tailors IP laws to the IC industry is rarely seen. This is a niche area in which books are seriously lacking.

The publication of this book serves to address the shortage. It was written by reputable university professors and IP practitioners with a wealth of expertise and experience in IP for ICs. Unlike other books that try to cover as broad a discipline as possible, this book is tailored specifically for the IC industry. The technical aspects, which include semiconductor and IC technology as well as law aspects (IP rights) are covered in the book. Most importantly, interesting examples and case studies related to IP for the electronics engineering community are also included in the book. In the context of this book, the phrase "Integrated Circuit (IC)" is often used interchangeably with the words "semiconductor" and "electronics."

This book is meant to be a complete handbook for anyone who is interested in the topic of IP for ICs. Furthermore, since the contents of this book embrace a cross-disciplinary study at the intersection of the two fields, it was written in a language that is reader-friendly even to a layperson. This is important in view of the fact that the majority of the target audience is either not legally trained, or without any engineering background. The authors have also adopted a lively style of writing so that reading this book is a pleasure.

Full coverage on IP law in the context of electronics engineering is presented. A step-by-step guide to IP rights and protections, as in how to safeguard one's invention, how to stay away from the pitfalls of IP infringement, what form of IP to register for, when to apply, where to get assistance, etc., is provided. This book provides the reader with the IP knowledge to be more self-reliant, rather than solely depending on IP practitioners. The perception of "Leave the Law to the Lawyers" is no longer practical because IP attorneys may not be as knowledgeable as the inventor about the inventor's field. This may lead to cases whereby the inventor pays the full amount for patent filing but cannot get the full protection he expected to receive. However, this book by no means replaces the role of patent attorneys.

On the other hand, IP practitioners (patent attorneys, trademark agents, etc.), especially those dealing with cases of integrated circuits, may also derive valuable information from this book. This is because it is an IC book for IP practitioners, and conversely, it is also an IP book for IC engineers. With adequate technical contents of semiconductor and IC design explained in an interesting and non-technical manner as much as possible, the authors believe that IP practitioners will find themselves more appreciative of the inventions of their clients.

Who will Benefit from this Book?

Well, in case you are still hesitant about whether you should bring this book home, let us tell you precisely who our target audience is.

- IP novices
- IC/semiconductor/electronics engineering novices
- IP practitioners venturing into the technical field of IC/semiconductor/electronics engineering
- Scientists/researchers/engineers hungry for IP knowledge

Who is Currently Benefiting from this Book?

Intellectual Property for Integrated Circuits is presently used as a textbook for the course, Intellectual Property for Electronics Engineers, offered at Nanyang Technological University—Singapore.

Desired Reading Outcomes

After reading this book, you should:

- Have a clear understanding of IP and its importance in retaining and elevating the IC industry to greater heights.
- Have a clear understanding of the fundamentals of ICs.
- Be able to identify the type of IP that is most relevant to their IC invention.
- Be aware of the potential of IP as a tool for technology advancement and recognition as well as wealth generation.
- Know how to avoid possible infringements and respond to infringements that occur.
- Be more self-reliant rather than solely depending on IP attorneys. You should have developed the necessary skills to interact effectively with IP attorneys in the endeavor of protecting and benefiting from your own innovation.

Contents of the Book

The carefully crafted contents of the book begin with an introductory chapter (**Chapter 1**) on the global IC industry and the fast-growing IP market. It brings the reader back to the historical scenes where IC and IP were born. An overview of the semiconductor value chain and the evolving semiconductor business model is also provided in this chapter. In addition, the authors share in this chapter important aspects of IP and ethical boundaries such as how to circumvent ethical dilemmas or practice within ethical boundaries. The economic contributions of IP and IC from a global perspective are also described to accentuate the past and present records, and promising future potential of both industries.

Chapter 2 contains an analysis of the importance of IP rights for the IC industry. The concept of the knowledge-based economy (KBE) is first introduced. Following that, the synergistic interweaving of the KBE and the IC industry is illustrated in this section. In fact, it is the knowledge-intensive factor of the IC industry that makes IP a critical component in most, if not all, IC activities. The chapter further explores the business applications of IP rights for sole inventors and business entities. The eminent need for IP reuse by IC designers and the growth trend of IC patents are also provided. The chapter concludes with the critical importance of establishing good cooperative relationships between IC inventors and IP attorneys to achieve a win-win situation.

Chapter 3 provides a detailed description of the technical context of ICs. The chapter commences with a definition of ICs and is followed by an overview of the evolution of IC technology. The main goal of this chapter is to equip the reader with the essential and fundamental knowledge of ICs. Even though the field of ICs is in fact an immensely technical area of expertise, the authors strive to make the study of this chapter as interesting as possible by inspiring the audience with real-life situations or events familiar to anyone. Put simply, it is a nontechnical guide to the technical framework of ICs. Equipped with this knowledge of ICs, an IP attorney can communicate more effectively with an IC inventor, and in the process gain more confidence in comprehending IC inventions, thereby closing the technical gap between them.

Chapter 4 serves to enlist and introduce the various types of IP that are applicable to the specific area of ICs. These IP vehicles include patent, copyright, Layout-Designs of Integrated Circuits Act, industrial design, and trademark. The respective forms of IP differ significantly in the area they cover, the rights they confer, the procedure for obtaining them, and how they are maintained. Once again, interesting examples are used throughout the chapter to enhance understanding of the legal framework. Furthermore, the topics on computer software protection and firmware IP protection are also covered in two separate sections of the chapter. Finally, a section dedicated to describing how to interpret and extract useful information from a registered IC patent concludes the chapter.

Chapter 5 introduces the reader to the legislative mechanisms surrounding ownership and duration of IP rights in general, with an emphasis on protecting ICs. The chapter discusses patent ownership in terms of the owner of the invention and the duration for which the monopoly over that invention exists. Copyright is defined, and the types of works in which copyright can subsist are considered in turn. The chapter delves into the ownership of layout-designs and the criteria required to be met in order for rights to subsist. Lastly, consideration is given to ownership of trademarks. This chapter is intended to provide a broad overview of the various types of IP rights that are available in Singapore.

Chapter 6 deals with the most important aspect of IP and ICs—infringement. Consideration is given to the various types of IP infringement, the remedies available should a court of law find that infringement has occurred, and the ways of avoiding infringement, which is closely related to the measures one should take to personally safeguard their IP work. This chapter will be highly useful and practical for IC engineers, while at the same time assist IP practitioners in advising their clients. Here, the authors take the opportunity to provide many illustrative examples and case law to assist understanding.

Chapter 7 seeks to educate the reader about the procedures and formalities required to be met for IP rights to be protected. This chapter specifically explains the step-by-step process of filing an application for a patent or trademark in Singapore and internationally. The intention of this chapter is to provide the relevant information required by an IP owner seeking protection of his or her work, including IC engineers.

Chapter 8 takes a closer look at recent Singapore High Court and Court of Appeal decisions on matters relating to technology and IP law. Each case is thoroughly analyzed, and a conclusion on the broader implications of each judgment is provided. Particular attention is given to the Court's interpretation of relevant legislative provisions and how they are applied to the facts and circumstances in each case. The judgments have been selected specifically with the intention of demonstrating how disputes in relation to IP ownership, invalidation, and infringement are resolved.

About the Authors

KIAT SENG YEO received his B.Eng. (Hons) (Elect) in 1993 and Ph.D. (Elect. Eng.) in 1996 both from Nanyang Technological University—Singapore. He began his academic career as a lecturer in 1996 and was promoted to assistant professor, associate professor, and full professor in 1999, 2002, and 2009 respectively. He was Sub-Dean (Student Affairs) from 2001 to 2005. During this period, he held several concurrent appointments as Program Manager of the System-on-Chip flagship project, Coordinator of the Integrated Circuit Design research group, and Principal Investigator of the Integrated Circuit Technology research group at NTU. He is currently a board member of Microelectronics IC Design and Systems Association of Singapore (MIDAS), a member of the Advisory Committee of the Centre for Science Research & Talent Development of Hwa Chong Institution, Chairman of the Advisory Committee of Dazhong Primary School, and consultant/advisor to several statutory boards and multinational corporations in the areas of semiconductor devices, electronics, and integrated circuit design. Professor Yeo is the General Chair and General Co-Chair of the 2009 and 2007 *International Symposium on Integrated Circuits,* respectively.

Professor Yeo is currently Head of Division of Circuits and Systems and Interim Director of the new IC Design Centre of Excellence at NTU. His research interests include device characterization and modeling, RF IC design, and low-voltage low-power IC design. He has authored four books: *Design of CMOS RF Integrated Circuits and Systems* (World Scientific Publishing, International Edition, 2009); *Low-Voltage, Low-Power VLSI Subsystems* (McGraw-Hill, New York, International Edition, 2005); *Low-Voltage Low-Power Digital BiCMOS Circuits: Circuit Design, Comparative Study, and Sensitivity Analysis* (Prentice Hall, NJ: Upper Saddle River, International Edition, 2000); and *CMOS/BiCMOS ULSI: Low-Voltage, Low-Power* (Prentice Hall, NJ: Upper Saddle River, International Edition, 2002). Professor Yeo has filed/granted 17 international patents and published over 300 articles on CMOS/BiCMOS technology and integrated circuit design in leading technical journals and conferences worldwide. He is a technical reviewer for several prestigious international journals. Professor Yeo was awarded the Public Administration Medal (Bronze) on National Day 2009 by the President of the Republic of Singapore and received the Nanyang Alumni Achievement Award in 2009.

MR. NG KIM TEAN, an engineer-turned-lawyer, holds a Bachelor of Engineering in Electronics and Electrical Engineering from the National University of Singapore and a Bachelor of Law (Honours) degree from the University of London. He is also qualified as a Barrister-at-Law (Middle Temple, London), an Advocate and Solicitor of the Supreme Court of Singapore, and a Singapore Registered Patent Agent. Mr. Ng is the Singapore Representative, Patent Committee of Asian Patent Attorneys Association; a Member of the Honourable Society of the Middle Temple, London; the Singapore Academy of Law; Singapore Inventor's Development Association; and Association of Singapore Patent Agents.

Mr. Ng is founder and Chairman of Nanyang Law LLC and was one of the first in Singapore to be registered as a Singapore Registered Patent Agent. His primary area of practice is intellectual property law and his many years of hands-on experience have exposed him to all aspects of prosecution and management of intellectual property portfolios including patents, trademarks, registered designs, domain names, trade secret, and confidential information in Singapore and various foreign countries. With his strong technology background, Mr. Ng provides effective guidance in the field of technology licensing and research collaborations.

Mr. Ng is also an independent director of Jade Technologies Holdings Ltd which is public-listed on Catalist, Singapore Exchange Ltd. He is the Chairman of its Nominating and Corporate Governance Committee and a member of its Audit Committee and Remuneration Committee. He is a prolific speaker, and he also regularly authors articles, publications, and books. One of his notable book publications is *International Patent Law–Winning Legal Strategies for Registration, Litigation & Other Intricacies of Patent Law in All Major Markets* (Aspatore, Inc. 2004). He is a two-term President of the Singapore Inventors' Development Association; an Advisor to Singapore Management University, Business Innovations Generator; a Part-time Lecturer at Nanyang Technological University; and an Intellectual Property Expertise Observer of Beijing JZSC, China's Intellectual Property Expertise Centre of Judicature.

ZHI HUI KONG received her Ph.D. (Elect. Eng.) from Nanyang Technological University (NTU)—Singapore in 2006. Dr. Kong had worked as a Research Engineer at the Institute for Infocomm Research (I^2R) and became a Project Officer and Research Fellow at NTU in the subsequent years. She was then promoted to become a Teaching Fellow and is currently a Visiting Assistant Professor with the School of Electrical and Electronic Engineering at NTU. Her research interests include digital/mixed-signal circuit designs for low-voltage low-power applications and circuit/architecture designs for the emerging probabilistic CMOS (PCMOS) technology.

Contributing as a co-Principal Investigator, Dr. Kong has been successfully issued a S$16 million research grant from the Singapore Economic Development Board (EDB) to train 150 IC designers for the semiconductor industry. In addition, she has also been awarded a highly competitive research fund from the National Research Foundation as a co-Principal Investigator amounting to about a quarter million dollars in 2009.

MS. DANG BEE YOKE, Tricia, holds a Master of Science in Polymer Science and Technology from the University of Manchester Institute of Science & Technology and a Master of Science in Intellectual Property Management, National University of Singapore. She also received a Graduate Certificate in Intellectual Property Law, National University of Singapore. Ms. Dang is an associate Member of the Association of Singapore Patent Agents and a member of the ASEAN Intellectual Property Association. She is also a member of the Singapore Inventor's Development Association.

Ms. Dang is currently the Deputy General Manager of Nanyang Law LLC. She works closely with the inventors on all issues related to the patentability of technologies. This includes conducting prior art searches, evaluating technologies, assisting in patent drafting, patent filing, and prosecution work in various jurisdictions throughout the region, managing and maintaining large patent portfolios, and conducting patent due diligence. Over the years, she has developed expertise in technology scanning, assessing technology trends, and technology and competitive intelligence.

Free value-added materials available from
the Download Resource Center at ***www.jrosspub.com***

At J. Ross Publishing we are committed to providing today's professional with practical, hands-on tools that enhance the learning experience and give readers an opportunity to apply what they have learned. That is why we offer free ancillary materials available for download on this book and all participating Web Added Value™ publications. These online resources may include interactive versions of material that appears in the book or supplemental templates, worksheets, models, plans, case studies, proposals, spreadsheets and assessment tools, among other things. Whenever you see the WAV™ symbol in any of our publications it means bonus materials accompany the book and are available from the Web Added Value™ Download Resource Center at www.jrosspub.com.

Downloads for *Intellectual Property for Integrated Circuits* include an online version of the in-chapter references with helpful links to governmental agencies and articles to further learning and research.

1

Introduction

Intellectual property (IP) and integrated circuits (ICs) both are red-hot revenue-making industries that have enjoyed remarkable success in the present technological era. They are the primary contributors responsible for causing human development and global economic growth to escalate to the next level. IP law protects intellectual innovations, and the IC industry is one such area that incessantly provides an abundant realm for innovation. Therefore, the chemistry between IP and IC, or more specifically a customized coverage of IP law tailored particularly for the IC industry, presents significant opportunities and is of great importance. Before proceeding, it is judicious first to introduce the background information related to the IP and IC industries to prepare the reader for more detailed discussions in the subsequent chapters.

Phenomenal Growth of the Global Semiconductor/Integrated Circuit Industry

From commonplace consumer products such as basic home appliances, computers, and communication devices, to powerful space rockets and even missile defense systems, semiconductors/ICs have become the nucleus component of practically all state-of-the-art electronic devices. The burgeoning use and relentless demand for increasingly more sophisticated applications with augmented functionalities have in turn spurred a phenomenal growth of the global semiconductor/IC industry. In fact, since its invention, there has been enormous development in the IC industry and the trend is still marching ahead aggressively. Let us now look at a brief introduction from the historical perspective of the IC industry. This is followed by three sections on the semiconductor value chain, the past and present manifestation of the semiconductor business model, and the industry's economic contribution and impact.

History of the Integrated Circuit Industry

The most important device that forms an IC is the transistor. It is widely accepted as the most significant discovery of the 20th century. We begin with the history of the transistor to set the scene for our discussion of the history of the IC industry. Before the emergence of transistors, vacuum tubes were commonplace. Vacuum tubes played a pivotal role in early day scientific discoveries and in the preliminary inception of electronic innovations, which underpin our contemporary electronic technology. However, engineers later realized that vacuum tubes had several limitations. They were big, bulky, slow, power-hungry, and unreliable. Hence, in the search for a replacement for vacuum tubes, two research scientists at Bell Telephone Laboratories (currently known as Bell Laboratories—a research organization of Alcatel-Lucent), John Bardeen and Walter Brattain discovered the first point-contact transistor in 1947 (see Figure 1-1). This important discovery resulted in U.S. Patent 02,524,035 issued on October 3, 1950.[1] Shortly after that William Shockley, who was the leader of the research team, developed the much more practical bipolar junction transistor in 1949 that led to another U.S. Patent 02,569,347 issued on September 25, 1951.[2] The junction transistor superseded the point-contact transistor in most applications by the mid-1950s.[3,4] The trio was then awarded the Nobel Prize in Physics in 1956 "for their researches on semiconductors and their discovery of the transistor effect."[5] (The name *transistor* was coined by John R. Pierce, a Bell Laboratories engineer.

At present, another family of transistor, Metal-Oxide Semiconductor Field-Effect Transistors (MOSFETs) are more power efficient and popular than their bipolar counterparts, as evidenced by their full-bloom exploitation in the semiconductor industry today. The original idea of the FETs was conceived in 1925 in a patent application by German scientist Julius Edgar Lilienfeld. He was awarded two U.S. patents for his contributions in FET-related inventions.[6,7] In 1934, a structure similar to the MOSFET, which was independently developed by Oskar Heil, was proposed in England. He was granted a British patent in 1935.[8] Insufficient knowledge of the materials, as well as gate stability problems, delayed the practical use of the device. The bipolar devices maintained their position at the top echelons of the digital technology arena until the late 1970s. A transitional change in this trend occurred in the 1980s when MOS technology eventually caught up and there was an intersection between the market shares of bipolar and MOS devices. More explanations of the evolution of IC technology are provided

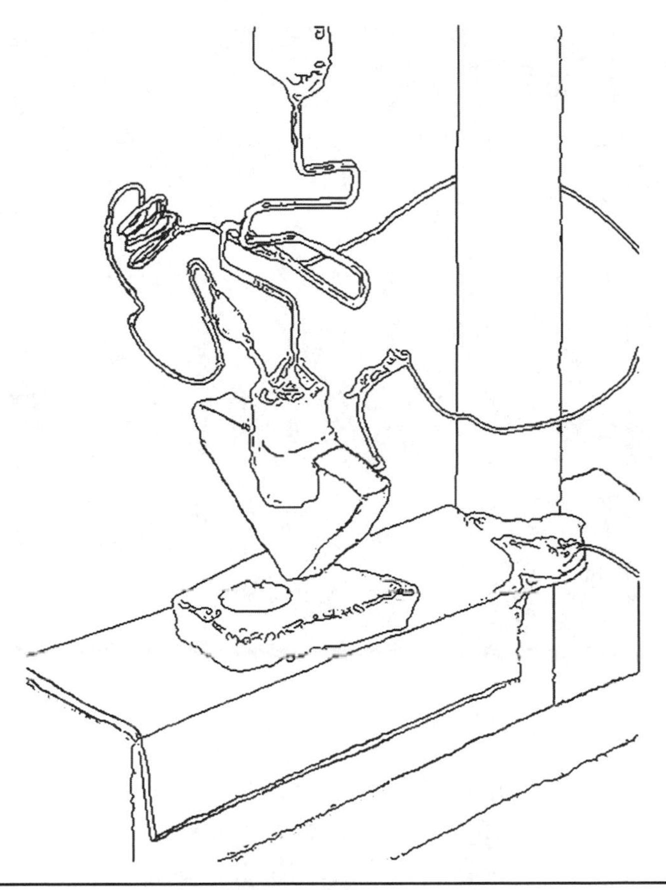

Figure 1-1 The first point-contact transistor

in Chapter 3. The invention of the transistor devices opened new doors to an unprecedented revolutionary age of the semiconductor industry. Thereafter, with the continuous advancement of semiconductor technology, the transistor incessantly became smaller, cheaper, faster, and more reliable.

A decade after the invention of transistors another breakthrough, the monolithic IC, was introduced by Jack Kilby from Texas Instruments (TI) in 1958.[9] The invention of the IC was motivated by the time-consuming, expensive, and, at that time, notoriously unreliable hand-soldering process of connecting hundreds or thousands of discrete elements. Photos of the first IC and Kilby are shown in Figure 1-2. Called an IC or simply a chip, it is a miniaturized electronic circuit comprised of a single transistor and other integrated components fabricated on semiconductor material.[10] The IC was filed for a patent in 1959 and granted U.S. Patent 3,138,743 in 1964.

Figure 1-2 The first integrated circuit and Jack Kilby (Courtesy of Texas Instruments)

The concept of ICs surfaced even earlier in 1952 when radar scientist Geoffrey Dummer first conceived the idea of the integrated circuit on May 7, 1952, whereby all electronic devices could be made as a single block.[11] He was thus dubbed the Prophet of the Integrated Circuit.[12] However, Geoffrey Dummer's attempt to build such a circuit in 1956 was unsuccessful. It was Jack Kilby who turned the potential into reality by inventing the IC. Kilby's tiny chip has since made history and has had a far-reaching impact in revolutionizing the world of electronics.[10] Currently—only some 50 years since Kilby's invention—ICs are omnipresent in nearly all areas of human endeavor. As TI so aptly put it, "The chip virtually created the modern computer industry, transforming yesterday's room-size machines into today's array of mainframes, minicomputers, and personal computers. The chip restructured communications, fostering a host of new ways for instant exchanges of information between people, businesses, and nations."[10] In 2000, the Nobel Prize in Physics was awarded to Jack Kilby in recognition of his effort in the invention of the IC.

The history of ICs does not seem to be complete without mention of another great man, Robert Noyce, who was known as the Mayor of Silicon Valley. He was the co-founder of Intel Corporation, which is currently one of the largest semiconductor companies in the world. In 1959, Robert Noyce conceived a similar idea to Kilby's. He was working at Fairchild Semiconductor when the idea of realizing a whole circuit on a single chip struck him. Noyce was acknowledged as having independently conceived the idea of the IC. An exceedingly detailed patent description of this invention was filed and later resulted in U.S. Patent 2,981,877 granted on April 25, 1961,[13] while Kilby's patent was still under consideration. Hence, Noyce was legally recognized and credited as the inventor of the IC along with Kilby. However, Noyce never won a Nobel Prize for his work on the IC as did his co-inventor—he died a decade before the award was given.

Since the 1960s, the progress of the semiconductor industry has continued to soar at a tremendously fast pace. In 1965, a pioneer of the IC-related era and cofounder of the Intel Corporation, Gordon Moore, then a young engineer at Fairchild Semiconductor, published an article in *Electronics Magazine,* "Cramming More Components Onto Integrated Circuits." This article was reprinted in the Proceedings of the IEEE in 1998.[14] Moore reportedly penned: "The complexity for minimum component cost has increased at a rate of roughly a factor of two per year." In other words, the number of components per IC was experiencing a twofold increase approximately every 12 months with a commensurate reduction in cost per component.[15] This observation later became the well-known Moore's Law. In 1970, Moore predicted that the cycle time of doubled complexity would slow down to two years, and amazingly, this law continues to hold.

Semiconductor Value Chain

The semiconductor/IC industry is formed by a spectrum of companies complementing one another in a sustainable way. The beauty of diversity in the semiconductor industry is that it leverages synergies among fabless and fab-lite companies, assembly and test (A&T) facilities, electronic design automation (EDA) houses, and wafer foundries. According to the Global Semiconductor Alliance (GSA) (previously named Fabless Semiconductor Association, [FSA]), fabless refers to the business methodology of outsourcing the manufacturing of silicon wafers, which has become the preferred business model in the semiconductor industry. They are also called pure-play IC design houses. Meanwhile, fab-lite is defined as integrated device manufacturers (IDMs) or vertically integrated companies with a strategy bent toward utilizing a fabless approach. A fab-lite company outsources less than 75 percent of its wafer manufacturing and includes research and development (R&D) as one of its core businesses. Different players in the industry leverage on and bolster each other while promoting a realm for technological innovation, scientific advancement, engineering excellence, and vigorous competition. In turn, an integrated enterprise ecosystem evolves comprised of players in different clusters who locate in close proximity to the design and production facilities, hence reinforcing each other's growth and contributing to a vibrant semiconductor value chain.

A typical semiconductor value chain comprises the silicon refinement plant, pure-play IC design house, wafer fabrication foundry, IDM, A&T facility, and other support industries, as shown in Figure 1-3. In general, the final product of the semiconductor value chain, semiconductor chips, is not sold directly to the end user. Instead, it is more often used as a key intermediate input in a wide range of manufacturing industries for developing electronics systems such as communication devices, electrical home appliances, computer hardware, automobiles, and even spaceships. As a result, the semiconductor industry has become a crucial bottleneck for an increasingly diversified combination of industries.

IC design houses present an instrumental impact in boosting and strengthening the entire semiconductor ecosystem. IC design forms one of the most sophisticated stages of the value chain, requiring highly skilled IC design talents. EDA tools and software packages (including some standard and commonly used IPs from the various EDA companies) are prerequisites for IC design. Some of the well-known EDA companies are Cadence, Synopsys, Mentor Graphics, and ARM. The R&D quality essentially determines a manufacturer's brand identity and brand image, because even leading-edge electronic commodities become obsolete rather quickly due to the increasingly fiery global competition leading to incessant rapid technological advancement. Being fabless, IC design houses are talent-intensive instead

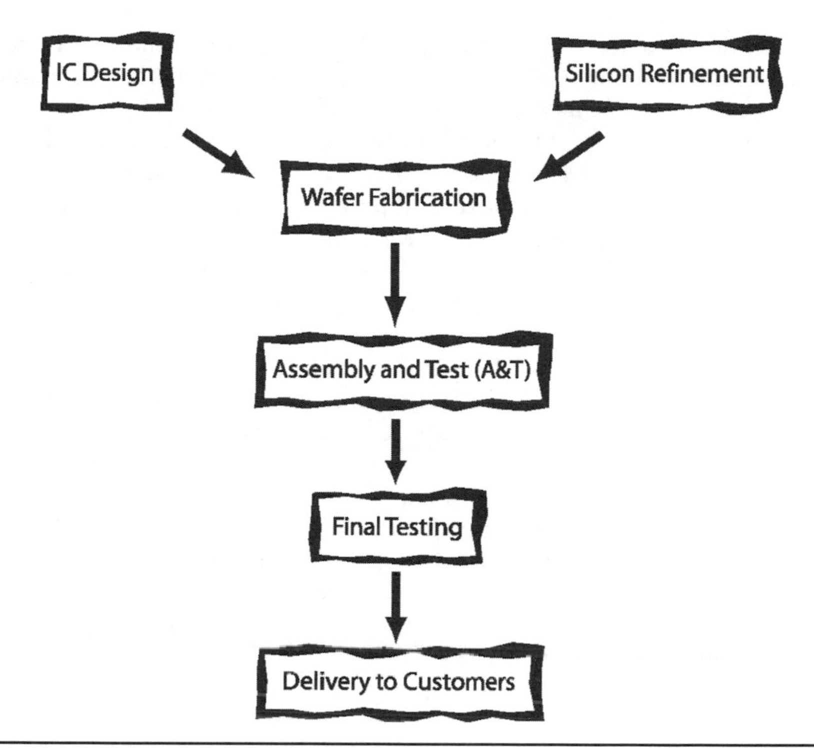

Figure 1-3 A typical semiconductor value chain

of labor-intensive and are free from the exceptionally high fixed cost of internal fabs resources. Besides incurring fewer fixed asset investments compared to foundries and wafer fabs, the IC design industry drives the demand for such foundries and wafer fabs, thus contributing to the virtuous cycle of semiconductor chips and the industry.

The wafer, which is made of silicon, is the fundamental physical unit used in the processing of chips. Although the raw material for silicon, silicon dioxide, is readily available in natural substances on the earth, including beach sand, it has to be refined through an exhaustive purification process. The silicon refinement stage is usually performed by firms that specialize in wafer production, whereby the level of purity is often refined to the desired extreme range of 99.9999999 percent. The pure silicon is then grown into cylindrical ingots and subsequently sliced into extremely thin discs, known as silicon wafers, using a diamond saw. Currently, the industry's largest wafer size has a diameter of 12 inches (300 mm). Upon the completion of a series of processes, such as polishing and cleaning to remove the saw marks and damage caused by slicing, the wafers, with a mirror-like finish, are then shipped out to the wafer foundries for IC chip fabrication.

Wafer fabrication is the commonly used term that refers to the entire procedure of constructing ICs on the wafer. In contrast to IC design companies, which are more agile in adapting to a fast changing market environment, the semiconductor wafer fabrication plants are less resilient in weathering the inherently cyclical global downturn of the electronic industry. This is because they are often burdened by the massive capital involved and also the high depreciation costs of the fabrication plants. In the wafer foundries, the circuit pattern of the chip, which is based on the circuit layout from the R&D stage, is reproduced many times and formed on each wafer through a long series of sophisticated and repetitive processes. Depending on the complexity of the IC, the masking/etching/deposition/doping steps are replicated numerous times until the circuit and its connections are completed. The fabrication process has to take place in an environmentally controlled clean room, as even the slightest form of contamination poses the risk of completely jeopardizing the intended circuit functionality. Upon completion of the fabrication steps, each device, or die, is electrically tested based on a test structure on the die by using automated test equipment (ATE), in a process known as device testing. The proportion of good dies, which are found to be functioning correctly on the wafer, is known as the *yield*. Then, the identical multilayer die grids on the wafers are separated into individual miniaturized integrated circuits using precision saws. This is called the dicing process and is not necessarily performed in wafer foundries. The good dies are now ready to be sent for package assembly.

The next participant in the value chain is naturally the A&T facility. In the chip assembly stage, the dies are incorporated into a package—also known as a chip carrier—with a substrate made of ceramic, organic, or plastic material. This process is also commonly referred to as die encapsulation. With the advancement of the assembly technology, a myriad of packages are available on the market and the selection of packages may differ according to the intended utility of the IC chip. The main purpose of this stage is threefold: (1) to protect the reliability of the IC from the surrounding environment, (2) to provide the IC with a structure to connect to the external world, and (3) to optimize the operation of the device. Basically, the two major steps involved in the assembly process are die attach and wire bonding. Die attach is a process whereby the chip is firmly mounted onto the die cavity of the support structure, e.g., the leadframe of the package; whereas wire bonding provides electrical interconnections between the pads of the die and the external leads of the package by using bonding wires made of gold, aluminum, or copper, which are thinner than a human hair. Subsequently, the packages are attached to a thin board, the printed circuit board (PCB), which is often used to mechanically support and electrically connect the package with other electronic components using conductive traces etched from copper sheets and printed onto a nonconductive substrate usually made of fiberglass. This PCB then directly serves as *input* for a multitude of industrial applications such as wireless communication, automotive designs, consumer electronics, and computer hardware.

The chip is ready for utilization once it is assembled. However, the chip must again undergo a variety of electrical tests before being shipped to customers. This is because the assembled chip may not work at all times due to the many possible uncertainties and imperfections that occur. For instance, the die may crack during assembly or the die-to-pin interconnect operation may not be carried out correctly (wire bonds not properly connected or not connected at all). Either of these is detrimental to the functionality of the chip.

Semiconductor Business Model

The semiconductor business model has undergone changes and evolutions since its early days. This, to a large extent, was motivated by the staggering development in the semiconductor manufacturing industry and the emergence of high-integration system-level designs. In the 1970s,[16] the semiconductor industry adopted a business model widely known as a *vertical integration* model, whereby a diversity of business operations within its value chain, IC design, fabrication, assembly & test (A&T), packaging, electronic design automation (EDA), were all integrated and came under the same umbrella (see Figure 1-4). During that period of time, the application

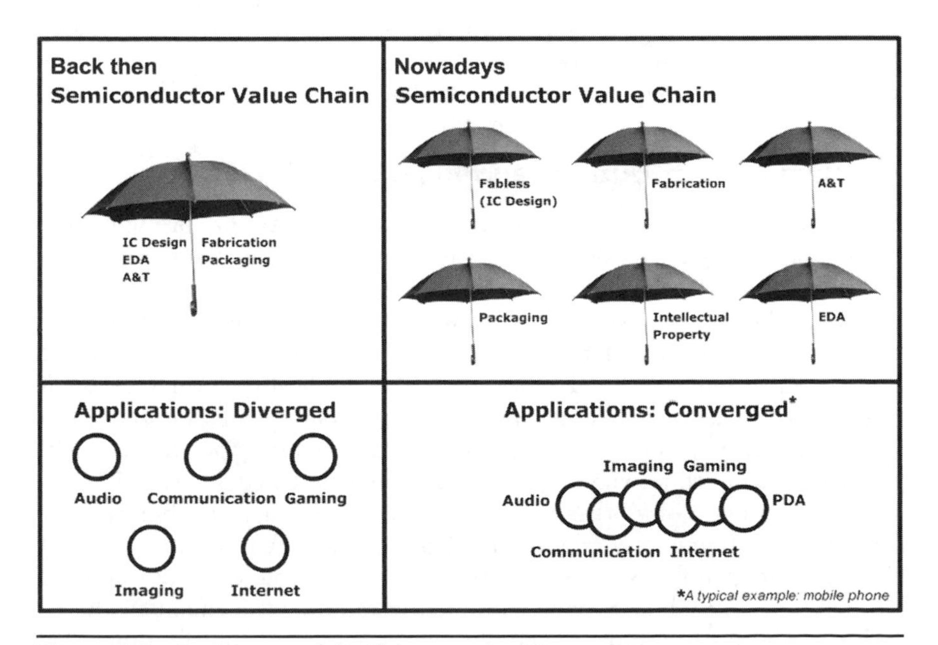

Figure 1-4 Business models of the past and the present

domain was distinct and less sophisticated, and consumers paid for products that served specific functions offered by the suppliers with bare-minimum features. For example, a mobile phone during that era was solely used for communication. It had no camera, audio, internet, nor gaming functions.

In recent years, the semiconductor industry has witnessed decomposition and specialization of its value chain into a business model known as a *vertical disintegration* model, and its business operations no longer come under the same umbrella.[17,18] Furthermore, semiconductor intellectual property (SIP), the reusable entity indispensable for contemporary IC design, surfaces and transforms the traditional look of the semiconductor value chain. This trend is largely fueled by the rapid advancement of technology, the extraordinary design complexity, and an exceptionally high IC development cost. Moreover, an apparent convergence and generalization of the application domain has also been observed. The boundaries among varying application domains have become fairly vague, and consumers currently enjoy the most integrated solutions with wide-ranging utilities.

In summary, the semiconductor industry value chain has moved from integrated to disintegrated, whereas its application domain has evolved from diverged to converged, albeit not coincidentally. Figure 1-4 illustrates the transition by using the mobile phone as a typical example. Previously, mobile phones were used exclusively as a communication device. Today, it is no longer just a phone. Apart from its conventional use, mobile phones serve a wide range of purposes, e.g., as an imaging device (camera/video), audio device (MP3 player), location positioning device (global positioning system [GPS]), and handheld computer (personal digital assistant [PDA]), all of which were historically streamlined into diverged application domains.

Economic Contributions and Impact on Society

The proliferation of ICs in the present technological era has brought forth revolutionary digital modernization that has ultimately transformed the history and lifestyle of humankind. It has bestowed upon us an everlasting change in the way we live, learn, play, work, and communicate. The IC is at the heart of recent developments in communication, automobiles, health and safety, the internet, and other numerous applications, while realizing never-imaginable breakthroughs in all facets of society. For example, a vast improvement from how things were in the past, whereby people had to be hooked up to fixed power supply points whenever using an electronic device, the evolution of IC technology has made portability possible, both for work (e.g., laptops and wireless technology) and leisure (e.g., MP3 players). Furthermore, people today would still be carrying around bulky waterbottle-like *big-boss* mobile phones without the rapid advancement in the IC industry. There are abundant examples of the significant contribution of the

IC industry to the lifestyle of humankind in the past and the present, and many more to be witnessed in the future. All these achievements would have remained an unrealized dream without ICs. Apart from the aforementioned, there are many more unsolved *mysteries* awaiting IC designers to decipher, develop, and subsequently bring further advancements.

The incessant escalation of IC functionality and the huge market demand for electronic products with a high level of sophistication has propelled the overall semiconductor market to rise tremendously within four decades from US$800 million in 1960 during its fledgling stage, to US$204 billion in 2000.[19] A graphical illustration of the worldwide semiconductor revenue over a 50-year timeframe is shown in Figure 1-5. The solid line indicates the actual revenue per annum, whereas the diamond-shaped symbols denote the revenue forecast by the Semiconductor Industry Association (SIA).[20] As can be seen from the graph, following the global economic downturn in 2001, which resulted in a massive plunge in semiconductor revenue, the industry was revived within a short span of four years with revenue at US$213 billion in 2004. A new record of semiconductor product global sales was set in 2005 at US$227.5 billion, and this record was swiftly overtaken in the following year and renewed at US$247.7 billion in 2006.[21,22] Today, the electronic industry has emerged to become the world's largest industry. In fact, the phenomenal growth of the semiconductor industry

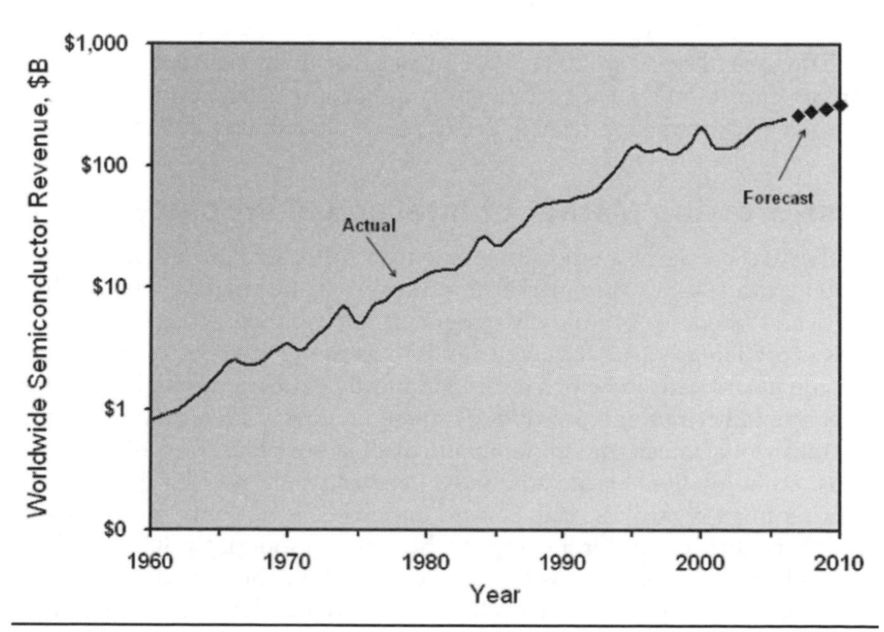

Figure 1-5 Worldwide semiconductor revenue and forecast (Source: Semiconductor Industry Association)

has evolved into one of the critical economic foundations that has rendered an immense impact to the global economy. The SIA forecast released in November 2007, projects the semiconductor industry to experience a continued growth to US$321.5 billion dollars in 2010.

For two decades, most of the worldwide output of the semiconductor industry has been used in the personal computer (PC) sector. However, in today's mass consumer market, the mobile electronic appliances sector has surpassed the PC sector in terms of semiconductor revenue.[23,24] Portable electronic products, ranging from small handheld personal communication devices such as pagers and cellular phones, to larger and more sophisticated multimedia-based applications such as laptop and notebook computers, have enjoyed remarkable success. Many countries, including the United States, Japan, Taiwan, South Korea, and Singapore have embarked on the formidable race of the semiconductor sector to vie for a share of the burgeoning industry. In the meantime, the IC industry in these countries is under immense pressure with more newcomers marching onto the scene. An emerging competitor is Dubai, with its economy traditionally dominated by petroleum. On October 12, 2002, Dubai Silicon Oasis (DSO) was launched. Investments in the high-tech design and manufacturing industry at the 7.2 million square meter technology park are estimated to exceed US$10 billion over a period of 20 years.[25] Furthermore, having emerged from its startup phase, China had more than 500 IC design houses in 2004, an evolutionary leap from just 68 in 1998, and its IC design talents mushroomed from about 1200 in 1998 to 20,000 by the end of 2004.[26,27] The fact that many new competitors are coming onto the scene conveys a clear message of the promising and overwhelming economic returns they see in the semiconductor industry.

Fast-Growing Market of Intellectual Property

Intellectual property, a critical ingredient to fostering innovation, is an important and fast-growing market in our present information age in which the world becomes profoundly dependent on technology and innovation. This is particularly true for countries such as the United States, the United Kingdom, and Taiwan, which are economically exuberant with knowledge-intensive industries and activities. In these countries, undermining IP protections would mean undermining one of the core pillars on which the nations' economy lies.[28] From a business perspective, IP is often the lifeblood of enterprises regardless of their sizes and fields. It not only forms a lucrative monetary stream for a company, but most importantly, it acts as both a shield and a weapon that is frequently powerful enough to determine the company's survival. Now, what is IP exactly? We ponder its definition in the following section. More specifically, we define semiconductor intellectual property (SIP), which is the main subject matter of our discussion.

Defining Intellectual Property

Property is defined as something that belongs to or is possessed by the owner. The first thing that usually comes to mind is something that is tangible, such as cash, jewelry, buildings, lands, cars, and equipment—something on which owners can physically lay their hands. However, the development of human intellect and the rapid advancement in technology, whether in the area of engineering, biotechnology, or humanities, has given birth to another form of property that is intangible—intellectual creation.

As its name implies, intellectual creation refers to creative endeavors arising from one's intellectual effort. Contrary to tangible or physical properties, intangible intellectual creation has essentially no material existence. In other words, it can neither be seen nor touched. It includes scientific inventions and discoveries; literary, musical, and artistic productions; symbols, names, images, and designs used in commerce; or computer software, technological know-how, business methodology, and integrated circuit topography. Ever wondered how such a brainchild is protected? Or, rather, is there even any possible form of protection for them? The answer is yes. Much like the owners of physical goods, intellectually gifted individuals also have the right to prohibit the unlawful deployment or exploitation of their creative works, albeit with distinct approaches. Whereas owners of tangible assets can safeguard their properties rather easily using safety measures of a physical nature, locks and fences are infeasible for innovators, authors, software programmers, artists, and others to protect their creative work.

Consequently, IP is the legal framework that serves this purpose; it legalizes the ownership of intellectual inventions and classifies them into various forms of IP that include patent, copyright, trademark, or layout design for integrated circuits, to name a few. In the context of the semiconductor industry, rapidly rising chip complexity presents an enormous market opportunity for semiconductor IP. Semiconductor IP (also known as integrated circuit IP) is interpreted as predesigned and well-tested blocks of circuits for utilization in designing complete semiconductor devices. These complex devices that usually make use of substantial volume of reusable semiconductor IP typically include system-on-chip (SoC), application-specific integrated circuits (ASIC), field programmable gate array (FPGA), and embedded systems. Instead of reinventing the wheel, the integration and sometimes repeated use of semiconductor IP components dramatically reduces the design complexity and development risk, thus accelerating time-to-market and enhancing design reliability while optimizing design cost. A detailed discussion on the various mechanisms of IP is presented in Chapter 3.

As in the case of all property, if you own an IP you can lease it to another party. You can also license it or even sell it whenever you like. However, just like the owner of a house who needs to be protected should his ownership be challenged, the rightful owner of an IP also needs protection should someone use his invention without being granted prior permission. This is where IP law comes into the picture. In fact, IP law is the field of law that gives the IP holder the exclusive rights to benefit from the results of their own creativity, and, at the same time, to control and forbid unauthorized parties from reaping the harvest of the owner's intense labor. In this way, the owner's intellectual efforts are rewarded as well as protecting the investment of time and the financial resources. This, in turn, promotes human creativity, benefits the public, and further encourages more innovation, research, and development to rejuvenate future global economic growth.

History of IP Protection

The origin of the legal protection of IP dates as far back as the 15th century. IP law has evolved into its present state over a timespan of nearly 600 years. The earliest known IP was in the form of a patent granted to the inventor Filippo Brunelleschi. It took place in the Republic of Florence in Italy on June 19, 1421.[29] It was an exclusive privilege conferred on Brunelleschi for a method of effective loading and transporting heavy merchandise on boats up and down the Arno River in Florence. Brunelleschi was recognized as the first great architect in the Italian Renaissance. It was when he tried to transport heavy stones across the Arno River to the city to build the massive Florentine cathedral dome that this new idea dawned on him. Before the patent was awarded to him, Brunelleschi refused to divulge his invention to the public for fear that it would be used without his consent. Subsequently, in 1421, the Gentlemen of the Works took up his case with the Lords of the Council of Florence to grant the exclusive right to Brunelleschi to monopolize the use of his invention on the waters of Florence for a period of three years. A peculiar voting method, using black and white beans, was used by the Gentlemen of Works in the decision-making process on whether to take up the case with the Council on behalf of Brunelleschi. The voting was met with overwhelming approval—218 black beans were cast in favor versus seven white beans against recommending the privilege.[30]

In the same century, the evolution of the patent also occurred in England. On April 3, 1449, King Henry VI awarded a patent to Flemish-born John of Utynam for his innovative method of stained glass manufacturing, an art which was previously unknown to the English. With the issuance of the patent, the King commanded that none of his subjects could use the stained glass arts for a period of twenty years without John's permission. In return, John was required to impart his skills to the King's subjects not only in the art of making stained glass but also many other arts never before practiced in the kingdom. Subsequently, John's invention was installed in the new educational institutions of Eton College as well as King's College at Cambridge.

Two centuries later, the United States had its first patent granted to Samuel Hopkins of Vermont on July 31, 1790, for his discovery of an enhanced method of producing potash and pearl ash with a new apparatus and process. Potash, or its more refined version, pearl ash, was a major industrial chemical used as an essential ingredient in the manufacture of soap, fertilizer, glass, and gunpowder. Since this was a chemistry-related invention, no prototype was needed for the patent application. Possibly it was due to this reason that Hopkin's application was filed much quicker than mechanically-related inventions. The patent was awarded for a term of fourteen years.

IP and Ethical Boundaries

Protection of IP and the enforcement of IP rights are crucial determinants of success for most technology-based enterprises. However, experience tells us that even with the existence of IP law, a flagrant breach or violation of the law occurs and has reached an alarming scale in the global arena. For example, despite the fact that many countries have established laws to guard against software piracy and unauthorized content exploitation, software piracy activities are still resulting in record global economic losses at nearly US$40 billion in 2006, according to a recent survey by BSA-IDC Global Software Piracy Study.[31] The study also unveiled that as much as 35 percent of the software installed on personal computers around the world was acquired through illegitimate channels. This serves to lend credence to the charge that the unethical use of intellectual creations is pervasive.[32]

As owners of IP (individuals, employers, or companies) or even as personnel granted access to sensitive information (employees or business partners), many times there are situations in which we are overwhelmed with ethical dilemmas in our attempt to protect the confidentiality of IP. Akin to a wealthy man (IP owner) who possesses ample cash and valuables (IPs) in his house, the dilemma is to what extent he decides to check on his servants (personnel given access to IP) to prevent theft of his valuables. On the other hand, his servants face the dilemma of resisting the temptation of stealing the valuables. Given this situation, the key question is: Where do we draw the line as far as ethical boundaries are concerned?

A company can implement safety measures with as many boundaries as possible to prevent confidential information from being compromised. Some of the commonly executable practices are e-mail monitoring, checking employees' belongings when entering and leaving company premises, and requiring employees to sign confidentiality covenants. The more uncommon ones include, but are not limited to, tapping of phone lines, installing closed-circuit television (CCTV) or voice recorders in meeting rooms or offices, discouraging employees from visiting a town in which a competitor is located, and making it compulsory for employees to disclose the companies for which their family members work. While the company may intend to carry out as many measures as possible, both common and uncommon, to protect their IPs to the fullest extent, employees may feel that their privacy is violated. On the other hand, if the company does not implement sufficient measures, it will find itself in a vulnerable position if employees deliberately misuse IP. This will ultimately lead to the leakage of confidential information and the company will lose a substantial amount of earnings, both present and future, not to mention its hard-earned reputation. Hence, in this case, who is right and who is wrong—the company or the employees? Is the implementation of the aforementioned protective measures exceeding ethical boundaries? All these are open-ended and vastly debatable issues.

Now, let us switch our attention to the nonowners of IP, (i.e., employees of a company or users of intellectual works). The preventive protocol implemented by the company will be effective to potentially alleviate unintentional leakage of confidential information by employees. However, if the leakage is premeditated, no amount of preventive measures can totally thwart this problem. In the same manner, it is nearly impossible to prevent them from using the nongenuine version if they are bent on this intention. Hence, for those individuals who deliberately violate the law, the pertinent questions are: How many more preventive measures are needed on top of the existing ones? How extreme should the measures be? Where is the point of equilibrium in order for ethical boundaries and IP enforcement to be balanced without any feeling of discontentment to either party? Similar to the issue of drinking and driving, how many more law enforcement officers do we need? How many roadblocks need to be set up?

As mentioned before, the topic of ethics is always an open-ended discussion. However, we have a strong faith that education provides one of the answers for the resolution to the problem. We see education as one of the most effective channels for us to inculcate the right concept of ethics and its boundaries to the younger generation. When they are exposed to the proper ethics with regard to IP rights early, they will mature with the knowledge that IP is something to be respected and not to be exploited.

Economic Contributions of Semiconductor IP

Semiconductor IP surfaced as a real market almost a decade ago and recorded worldwide revenue of US$67.8 million in 1996. It has since proven itself as one of the fastest growing sectors in the semiconductor industry despite dissenting voices along the way. It soon became the cornerstone that fueled the recent development of highly sophisticated system level designs such as ASIC and SoC.

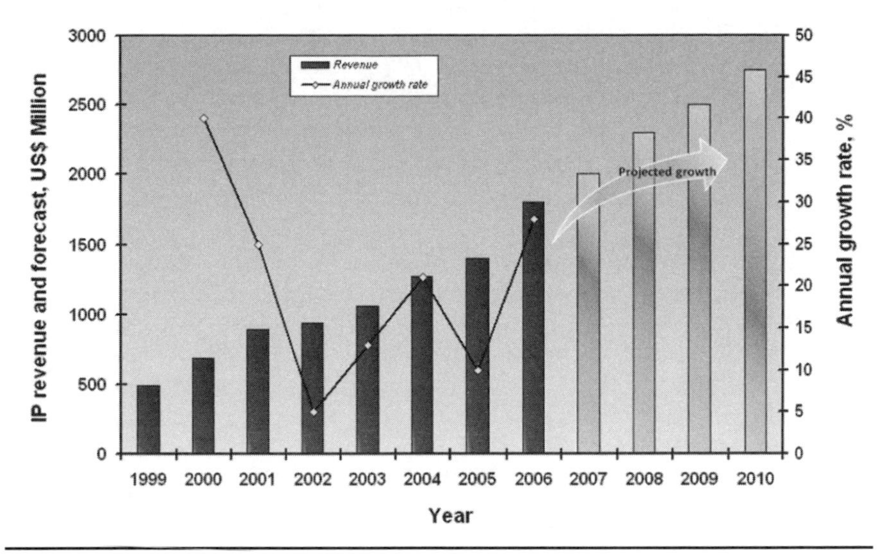

Figure 1-6 Worldwide silicon intellectual property revenue and forecast (Source: Gartner Dataquest)

At present, analysts remain optimistic regarding the future prospects of the silicon IP industry and predict that it will continue to show unrelenting advancement in years to come.[33] In August 2006, Gartner, the world's premier advisory company in information technology research and analyses, envisaged semiconductor IP revenue to exceed US$2.7 billion in 2010, an impressive twofold increase from US$1.4 billion in 2005 within a short span of five years.[34,35] This growth is projected based on the increasingly intense shift in customers' demand from simple IP (standard functions) to more complex IP with higher cost (more sophisticated specifications). An annual trend of the worldwide silicon IP actual market (1999–2006) and a four-year projection (2007–2010) is illustrated in Figure 1-6. The market demonstrated substantial escalation each year with some fluctuations in the growth rate. For instance, in the global economy recovery year of 2002, it grew only by a modest 5 percent with revenue of US$933.8 million. However, this growth rate outperformed that of the semiconductor chip industry, which grew by only 1.9 percent.

Some of the top players that contribute a sizeable share of the IP reuse market are multinational companies such as ARM, Rambus, and Synopsys. A high growth potential of the IP sector attracts potential venture capital investment. These investors are keen on identifying companies with innovative IP that will be sought after by a myriad of original equipment manufacturers (OEMs). These IPs will then be designed into potentially high volume products.[36]

2

Importance of Intellectual Property Rights for Integrated Circuits

Knowledge-Based Economy

The world economy went through salient structural transformation during the latter part of the 20th century. One of the important traits of this transformation is the apparently intensifying dominance of *knowledge* in almost all segments of global economic activities. In general, most major economies worldwide embarked on an agriculture-based economy with farmland and water being the requisite resources. The economies then went through significant transitions and reformations following the industrial revolution in which natural resources, machineries, and manpower were vital resources.

Today, with the onrush of technological forces, these advanced economies have evolved into their present state. It is essentially based on societies accumulating wisdom, knowledge, and information and is aptly named the knowledge-based economy (KBE). One of the earliest observations on knowledge being leveraged as an instrument for economic sustainability and performance was recorded in a time horizon that coincides closely with the invention of the IC in 1958. In 1962, in *The Production and Distribution of Knowledge in the United States*, economist Fritz Machlup examined the role of knowledge as a significant economic resource.[1] In 1996, the term knowledge-based economy was first coined by the Organization for Economic Co-operation and Development (OECD).[2] It is defined as economies that are directly based on the production, distribution, and use of knowledge and information. In fact, the IC industry is one of the economic segments that sowed the seeds for the KBE that was rapidly unfolding during that era, hence, shaping the global economy and environment we live in today.[3]

Knowledge is now the key motivator and power tool for the global pursuit of wealth, productivity, and economic prosperity. This resembles the inspirational story about King Solomon, who was renowned both for his impeccable wisdom and legendary wealth. Above all other worldly things, King Solomon humbly requested in his prayers wisdom, knowledge, and discernment. In return, King Solomon received more than what he had initially prayed for, and which far exceeded his imagination: unsurpassed riches, wealth, and honor all followed after he acquired knowledge and wisdom. He reigned over ancient Israel's Golden Age with profound peace and happiness and was widely acknowledged as the wisest and richest man who ever lived.[4]

In fact, knowledge resources such as innovation, talent, technical skills, and expertise, as well as IP, are becoming more crucial than traditionally important economic resources, including land, natural resources, or even human labor. The IC industry, which emerged and blossomed around the same time as knowledge was deemed an economic tool, has etched a niche in the global KBE given the fact that it is vastly brainpower-based and talent-intensive. In this regard, IP law is particularly imperative, as it is the legal foundation for the IC industry within the competitive KBE.

Now, let us elaborate in detail the importance of IP rights for the IC industry by using the IC design facility as an example. The process of designing a complex IC is an extraordinarily intricate, skill-demanding, laborious, and expensive process.

The career path of IC designers is inherently about invention and creation to improve the quality of human life. Indeed, any electronics gadget, be it as small as a car remote control to as large as a space rocket, requires IC chips to achieve its task. IC designers generating the IPs are analogous to gold miners prospecting for gold in a mine. Gold miners have to vigilantly navigate through the land before reaching the gold mine. Even after successfully harvesting the gold, they have to know how to safeguard and secure it to prevent the gold from being stolen. Similarly, in the midst of generating a new invention, IC designers must have knowledge of IP law to avoid the risk of exposing themselves to IP infringement litigation. Moreover, after successfully generating the IP, IC designers must know how to shield themselves with IP law. Generating an invention without protecting it is akin to mining the gold and then leaving the door open for people to steal, and the consequences can be disastrous.

Let us consider another analogy. Most of us have heard of the Golden Gate bridge. It is an internationally recognized quintessential symbol of San Francisco, California, and is widely acclaimed as one of the most beautiful bridges in the world. The construction of the suspension bridge cost more than US$26 million and took four years before its completion in 1937.[5] Now, imagine if someone were to copy the architectural design of the majestic Golden Gate. He or she could easily replicate the bridge design and build it somewhere in the world without having to kill too many brain cells or exert hardly any effort. This underlying concept is analogous to that of IC layout-design. The layout-design of an IC is, in fact, similar to the architectural design drawing of the Golden Gate. It involves a great deal of creativity and may require several to tens of millions of dollars over the course of several years to successfully develop a complex chip.[6] It is, however, exceptionally prone to competitive cloning and misappropriation. This is due to the fact that IC layout-design itself embodies the selection and configuration of electrical components and interconnections with precise geometry in order to realize the targeted electronic functions.[7] Thus, if the pirate copies the layout-design of an IC, he would be able to duplicate the IC that is capable of accomplishing the same function as the original chip, nearly effortlessly. This reflects the importance of safeguarding the layout-designs of ICs.

In a nutshell, the search for new and more innovative circuit designs would be undermined without legal protection via IP law. As a result, IC designers or inventors who are involved in the heavily knowledge-based IC industry should equip themselves with a substantial understanding of the numerous forms of intellectual property and their underlying laws and governing principles to efficiently protect their own inventions and, thus, profit from the ownership of intellectual property.

In the following section, we demonstrate how IP plays an indispensable role in the formation, sustainability, and development of IC-related businesses and enterprises.

Business Applications and Economic Contributions of IP

Along with the human race's inexhaustible ability in generating innovative ideas, IP is now everywhere around us. Simply consider something familiar and close to you: a mobile phone or an electronic gadget that many seem unable to live without. You may not even realize it, but it contains several forms of IP. The small brand logo engraved on it identifies the company that produces the mobile phone and is an IP known as a trademark; the microchip that gives wide-ranging functionalities and applications to the phone may be protected by a patent or the layout-designs of ICs act; the computer codes running on the microchip and the music stored in it can be protected by copyright; the chic plus the unique exterior is protected by another form of IP known as industrial design. As we systematically guide you through this material, you are exposed to brief glimpses of the different vehicles of IP for IC. Thus, the various forms of IP will not be a stranger to you when we discuss them in full in Chapter 4.

Beyond providing a fair remuneration for IC inventors and setting a foundation for governments to lure potential foreign investment and technology, IP for IC is also crucial to improving the lifestyle of the public. Without substantial IP support, advances in IC-related activities such as communication, automobiles, health care, and safety would not have come as far as they have. Additionally, IP protection creates IC business opportunities, drives new business models, and brings monetary rewards to individuals and enterprises. Mere ideas or IPs without a business built from them are meaningless. This is analogous to an idea for a movie. Having the storyline and the script are not sufficient: you have to make it into a movie to reach the masses. People are interested in the final product, not just the idea.

Consequently, like the bamboo and the river bank which are interdependent and derive benefits from each other's existence, IPs should be complemented with a strategic business approach to capitalize on the idea. Businesses, on the other hand, should exploit and leverage on the IPs they own for revenue generation and for sustaining a competitive edge. Taking it one step further, businesses should also proactively establish an IP management and protection strategy to ensure optimum profitability and operating efficiency. In the succeeding sections, we examine the business applications of IP and its contribution to economic progress, first from a single inventor's point of view and then from the perspective of a business or enterprise.

IP Protection for Single Inventors

Anyone can be an inventor—men, women, or even children—regardless of their profession, interest, or educational level. A single inventor, or a sole inventor, is an individual who conceives an ingenious idea or concept and further develops it into a feasible embodiment either on paper or using technical means—in legal terms, reduction to practice. A sole inventor can be someone who works for a company or organization; he or she can also be an independent or a private inventor not attached to any corporation. IP protection (the most prevalent being patent protection) for sole inventors is strategically important. It helps to pave the way to self advancement and career development, as well as opening new doors for future opportunities. There is also a prospect of gratifying wealth generation if the new product based on the IP receives enthusiastic response from customers.

In many cases, the novel idea is conceived by sole inventors during their daily work or routine activities while they themselves are actually engaged in a profession entirely foreign to the technology area in which the ideas were based. These are the independent inventors. For example, it could be a retiree who sparks off a new idea for a fishing lure design during one of his frequent fishing trips, or a homemaker who formulates a breakthrough metal polish as a result of her constant battle with tarnish and spots of corrosion. Through multiple corroborations and testing, our prudent retiree and homemaker inventors confirm the usefulness of their new inventions and are full of confidence that their inventions are set to make waves in the respective industries. Now the typical question is how to bring these ideas to the marketplace—into the hands of consumers—in the fastest possible way.

In general, the path to profiting from an invention runs the gamut from raw idea development, extensive market research, invention protection, transformation of idea into a usable product, and ultimately to marketing the invention in the form of a physical product. The entire course of transforming an idea into a product coupled with a successful marketing strategy is, in fact, a business endeavor with IP as a seed in the adventure. Since it is a largely multifaceted venture, it is unlikely that the sole inventor has sufficient time and the necessary expertise to do it alone. As such, the sole inventor normally proceeds by either licensing or selling the idea to a company already in the manufacturing business, or forming his or her own team. Progressing in either route, invention protection is still the primary factor to be considered. Hence, to prevent the loss or theft of ideas in the process of engaging prospective external support, the sole inventor should first *own* the idea. Ideas can be legally *owned* when they are protected by one of the many forms of IP. In other words, sole inventors can protect their inventions from unauthorized copying or use through IP, which can take shape in various forms: patent, copyright, trade secret, to name a few. In the context of the IC design industry, semiconductor IP generally appears in the form of patents, copyright, or mask work design.

With IP included in the business plan, it is much easier to attract financial support, because IP is viewed by investors and financiers as an asset. However, it is important to note that inventors should not rush to file a patent without conducting research to ensure the market worthiness of the idea. Patenting an invention is a lengthy process and may cost a significant amount of money. By adopting a lower-risk approach, smart inventors perform an invention assessment or market evaluation before dipping their toes into acquiring a patent.

Of course, the sole inventor can opt to set up his or her own small business based on the granted IP alone while aspiring to generate a handsome amount of revenue from the venture. Nevertheless, many, if not most of these start-up ventures fall apart due to hurdles such as an improper business plan, lack of experience, insufficient funds, and an uncompetitive marketing plan. Indeed, it is a challenging task with overwhelming obstacles and ramifications along the way to successfully develop and bring a new idea to the marketplace. For instance, the development cost can be prohibitive for a private inventor at the early stages. Therefore, it is not uncommon to come across creative ideas with great expansion potential that literally wither and die on the vine.[8] In view of that, sole inventors are encouraged to seek financial backing and expertise support from venture capitalists or business angels to increase the likelihood of a successful and enriching business endeavor.

IP Protection for Corporations

Not too long ago, physical assets were the primary determinant factor of a company's worth. They were also greatly responsible for positioning the company at a competitive edge in the worldwide marketplace. Over the years there has been a drastic change in this scenario. This has been as a result of the capability of IP turning ideas into valuable exclusive commodities that can often be merchandized by commercial and professional organizations of all sizes that include indigenous companies and multinational corporations (MNCs). Today, IP, an intangible asset in which proprietary knowledge exists, increasingly has taken over the place of physical assets and is now the core portfolio of a myriad of modern businesses and enterprises whose growth is fueled by knowledge and creativity.

The IC industry, which is to a large extent an area that increasingly provides a realm for innovation, is growing through this transition. An abundance of IPs are continuously being reused and at the same time new IPs are being generated in businesses within the integrated IC value chain. Frequently, IP appears to be the most valuable asset shown in the financial ledgers of indigenous IC companies or MNCs. For instance, semiconductor IP in the form of patents with underlying inventive ideas is crucial for developing sophisticated electronic products that are highly profitable. Smaller scale indigenous IC companies that are less financially equipped may also license their patented technologies to other companies that have the capacity to commercialize them, thereby bringing in remunerative royalty income for the business. Meanwhile, IP assets such as trademarks are much more than just appending a name or artwork to a product. Instead, it can help to promote the company's identity and distinguish its products and services from other competitors and foster a loyal clientele.

In view of the potentially significant economic prospects of IP to the growth and sustainability of a business, strategic IP management and its effective utilization is undoubtedly the key to greater marketability in the highly competitive IC market, be it domestically or internationally. Therefore, it is essential for upper management of IP-savvy firms to establish a keen view and look meticulously into the steps required for strengthening IP protection.

It makes intuitive sense that a company should know what it possesses in order to protect it. Therefore, the company should expend an effort to inventory a complete list and document its IP. Moreover, employees and researchers should also be reminded to reveal new discoveries at all times and be educated about the importance of IP to avoid inadvertent disclosure of IP information to outsiders. The company should also pay special attention to untapped intellectual works that warrant IP registration and protection—this could possibly be an unrealized opportunity that eventually turns out to be a value-add to the company. In addition, management should form an IP auditing team to periodically assess and determine the commercial feasibility and value of its existing bank of IP. By ascribing a value to IP, the plan for IP development and protection can be carved out more strategically so as to stay in line with the corporation's mission and goals.

Another measure that can lead to effective and timely IP protection is through manpower deployment. Unlike sole inventors, indigenous companies have employees and business partners who can act as their eyes and ears. This is exceedingly true for MNCs such as Intel, IBM, and Synopsys that have legions of employees all around the world. Companies can rely on their employees in the field to watch for and identify potential IP violation concerns, because management can only oversee a small portion of IP enforcement issues of the IC industry within a confined region. Thus, coupled with adequate training and comprehension of IP rights, employees' alertness and awareness of industry trends can aid substantially in monitoring the development and protection of the existing IP rights of the business.[9]

One also has to be aware that IP rights are usually territorial in nature. This signifies that IP rights can only be enforceable within the geographical boundaries of a country where protection has been registered or granted. For example, a patent acquired in Singapore may not necessarily be recognized and protected elsewhere in the world. Therefore, IP rights are to be sought for in all potential regions where the company intends to trade in order to enjoy exclusive IP protection in its export markets. IP rights in foreign territories can be dealt with through either national application (an application for each respective country), regional application (a single application for countries located within the same region), or international application (a single application for international protection).[9]

On top of IP protection, IC companies should also dedicate ample time and effort into managing, enforcing, and commercializing IP to attain the greatest possible commercial gains from its IP ownership. On the other hand, businesses should stay vigilant at all times to avoid infringing upon the IP rights owned by others. If a business has to use IP that belongs to others, then it should consider buying it or acquiring the rights to use it by taking a license or an assignment to prevent potential legalistic disputes that often involve consequential high-cost litigation.[10]

IP Reuse for Integrated Circuit Design

The unceasing escalation of silicon capacity spurred by the rapid progress of silicon manufacturing technologies allows IC and system designers to integrate increasingly sophisticated circuits and systems on a single IC (chip) of silicon—an integration platform widely known as system-on-chip (SoC), which is an evolution of the application-specific integrated circuits (ASIC) technology. However, with a substantial increase in system complexity, it becomes a severe challenge, in fact, an almost impossible one, for companies to continue their previous pattern of designing and developing products within a short time window. This incurs a precipitous lag in design productivity and, in turn, presents a critical impediment to the growth of the IC industry.

In an effort to circumvent this issue, semiconductor IP reuse was conceived and touted as one of the most prevalent solutions to bridging the broadening gap between silicon capacity and design productivity.[11,12] This is repeatedly accentuated in recent publications of the *International Technology Roadmap for Semiconductors* (ITRS) that states that IP reuse is one of the main factors that drives design productivity, and it promises to allow assembling SoCs of unprecedented complexity in a short time.[13] Moreover, the need for reusable IP continues to increase with growing design complexity.

As its name implies, IP reuse refers to reusing or directly incorporating an IP core, which is a predesigned and preverified building block or function protected by patent, layout-design of IC act, or copyright, in a new design.[14] The IP can be modified and tweaked to suit the specifications of a particular application. In many cases, appropriately reusing work that is already done in a complex design leads to higher efficiency, less arduousness, and, at the same time, provides more time to innovate.[15,16] IP reuse may refer to internal reuse of designs within a group of designers, within an organization, or externally from third-party IP vendors, albeit being conceptually identical.[17] Henceforth, the terms semiconductor IP reuse, integrated circuit (IC) IP reuse, or simply IP reuse, are used interchangeably.

The term IP reuse first entered the IC industry lexicon in the late 1980s when application-specific integrated circuits (ASIC) companies rushed to vie for a share in high-volume market sectors such as consumer electronics that were attractive money spinners by delivering application-specific standard products. One way to face the intense time-to-market pressure without reinventing the entire IC design wheel was by licensing microprocessors such as Scalable Processor Architecture (Sparc) and Microprocessor without Interlocked Pipeline Stages (MIPS) and build standard peripheral

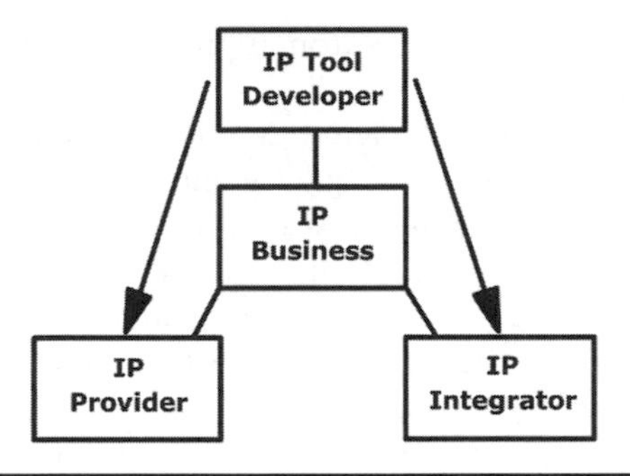

Figure 2-1 The IP business ecosystem

blocks around them.[18] Therefore, ASIC libraries can be considered as one of the pioneering design methodologies that encompass IP attributes to enhance design productivity. This gave rise to a new business strategy within the semiconductor industry, namely the IP business model.[11]

As shown in Figure 2-1, core players in the IP business ecosystem are the IP provider, the IP integrator, and the IP tool developer.[11] While IP providers such as ARM and Virage furnish IP blocks that include hard and soft cores, IP integrators are literally designers and application teams in fabless design houses, such as Broadcom, Qualcomm, and Xilinx. The reader also needs to be aware that a meager drop-and-stitch IP integration process does not warrant efficient IP reuse.[18] Instead, experience and technical know-how of the cadre of designers, coupled with the support from IP tool developers, are imperative to weave a conducive and seamless environment for IP reuse. IP tool developers or electronic design automation (EDA) companies such as Synopsys and Cadence are responsible for providing IP providers and in-

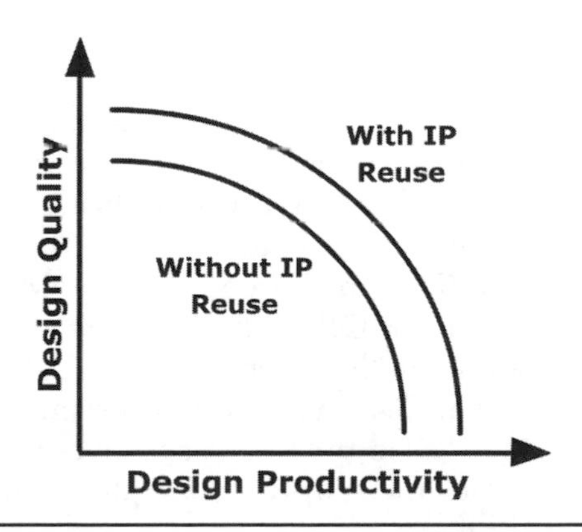

Figure 2-2 The importance of IP reuse on design quality and productivity

tegrators with innovative design automation software and tools to support the IP system integration.

The graphical illustration in Figure 2-2 shows the theoretical impact of successful IP reuse on the design quality and design productivity.[19] It adequately exemplifies that IP reuse is important not only to boost design productivity but also to salvage design quality. It also shows the trade-off between design productivity and design quality in general: as designers strive for higher design productivity, they usually have to resort to sacrificing design quality, and vice versa.

There are international organizations that help to foster the proliferation of IP reuse in the semiconductor industry. An important association widely perceived as the umbrella organization for IP and design reuse is the Virtual Socket Interface (VSI) Alliance which has been in operation since 1996.[20] It is an open organization inaugurated with joint efforts from leaders of all segments of the SoC industry including semiconductor, semiconductor IP, and EDA companies around the world and is devoted to the development of SoC, IP, and reuse standards to enhance the productivity of SoC design.[21]

Growth of Integrated Circuit Patents

The importance of IP to the entire IC industry can be exemplified by the explosive growth in the number of patents awarded to IC inventors. Among the many types of IP, patent is the most frequently sought for strategically important IC inventions because it provides the strongest legal protection to its holder. Its role is indispensable for the commercial viability and development of today's high-tech industries—the IC industry is undoubtedly one of them. Therefore, it is appropriate for us to treat patent as a representative form of IP to illustrate the growing importance of IP for IC.

The number of patents in the IC category has been increasing rapidly. This growth trend of IC patents is reflected in our patent analysis as shown in Figure 2-3, which is presented in five-year intervals from 1970 to the present (to June 2008). The search was conducted based on Scopus, which is presumably "the largest abstract and citation database of research literature and quality web sources" as claimed on its website.[22] The collection of patent records considered here are issued by five prominent patent offices in the world, the United States Patent Office, the World Intellectual Property Organization (WIPO), the European Patent Office, the UK Patent Office, and the Japanese Patent Office. Two keywords, semiconductor and integrated circuit, were used in the patent search.

As seen from the graph, there has been a substantial rise in the number of patents granted for ICs in the time range investigated. More specifically, having emerged from the start-up phase, the number of IC patents

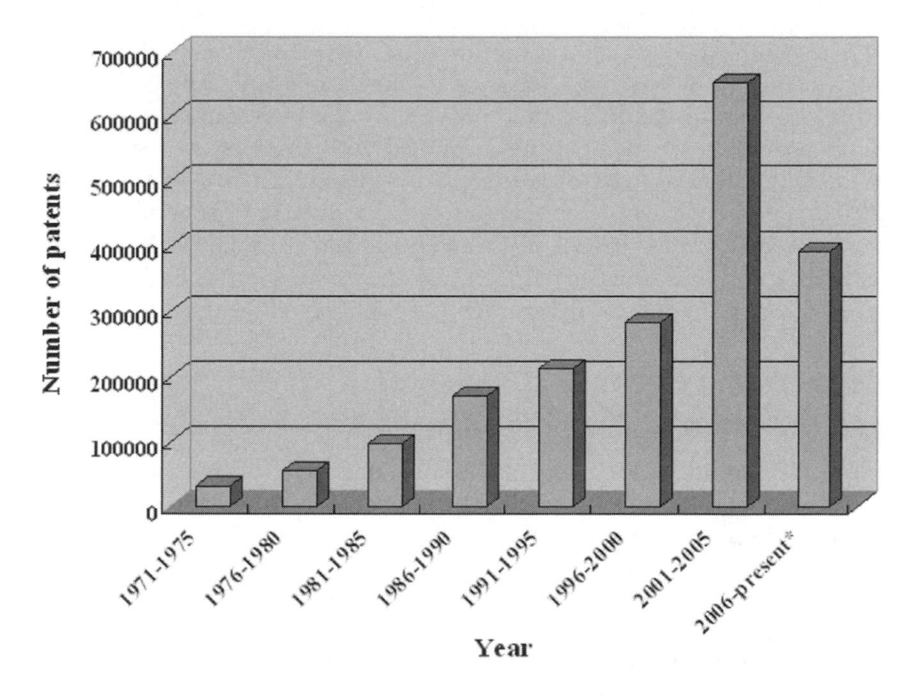

Figure 2-3 IC patents: A growing trend

granted was 32,624 within the time bracket of 1971 to 1975 inclusive. In the 1976–1980 timeframe, the number grew approximately two-fold to 56,690 with a record 42 percent growth. The patent numbers continued to increase somewhat drastically with 42 percent growth for another two terms over their preceding five-year time spans. During the transition from the 1986–1990 (170,775 patents issued) to 1991–1995 (212,871 patents) time spans, a relatively less aggressive increase in the number of patents was observed with a 20 percent growth rate. A similar rate of progress could be seen in 1996–2000 with a 25 percent steady increase in patent counts as compared to the preceding time interval. During recent years, i.e., the 2001 to 2005 timeframe, an abrupt rise of 57 percent was observed with 653,236 patents issued. This figure also shows that the patent counts in the 2001–2005 period has seen a dramatic growth of more than 20 times compared to the early years. As for the 2006 to the 2008 bracket, the patent search result suggested that there has been a subtotal of 391,270 patents being granted to IC inventors as of June 2008.

The patent analysis above suggests that there has been a rising trend in the number of patents related to IC inventions since the 1970s. This promising trend is perhaps made possible by the huge influx of emerging IC companies and related entities, coupled with the increased strength in IC-related patent activity of academic or research institutions throughout the years. Moreover, this may translate to an increase in the confidence level and awareness among IC engineers and inventors of the importance of obtaining patent (and therefore IP) protections to safeguard their intellectual creations as well as securing commercial monopoly and exploitation.

Cooperation between IC Inventor and IP Attorney: Win-Win Situation

Close cooperation and timely interaction between the inventor and the IP attorney (IP practitioner or IP agent) is crucial to secure a strong patent protection. If you are an inventor of IC-related products, you should equip yourself with the knowledge in the IP regime to be more self-reliant rather than solely depending on your IP attorney or IP agent. This is especially important if patent protection is the specific type of IP that is being sought. In this respect, the notion of *leave it to the lawyer* is clearly outdated. Patent attorneys may not be as knowledgeable as you who are the proud inventor of an IC-related creation in your particular field of expertise. Therefore, close cooperation and consistent interaction between the inventor and the patent attorney is essential especially in patent filing and prosecution. Failure to abide by this advice may lead to situations whereby the inventor paid the full amount for IP protection, especially in patent filing, but eventually is not granted the full protection that he was supposed to enjoy. In some critical cases, this could even land the entire business or company in dire straits. If this happens, the patent attorney in turn puts his or her reputation in jeopardy and suffers a great loss due to an eroding client base. The following paragraphs exemplify how inventors and patent attorneys can work together to obtain the strongest and most effective patent coverage.

Let us first illustrate the scenario with one example. As you might be aware, a preliminary or prior art search for a supposedly new invention has to be conducted to ensure its novelty before deciding on the feasibility of a patent application. Prior art includes all information available or existing prior to the particular invention for which patent protection is applied. Instead of relying totally on your patent attorney to search for relevant prior art, you should also play an important part in this process because you are well-versed in the specific area of your invention. In view of this, you should prepare yourself with the know-how to search for and interpret prior arts.

In fact, you should conduct your own prior art search before seeking legal advice because there is no point to proceed if the results of the prior art search clearly indicate that your invention is already known to the public. This will definitely save you precious time and money. However, in most cases, it is not that straightforward to determine if your invention is new in the eyes of the law. You need to compare your invention against the other prior art and assess their similarities and differences. This discerning capability is normally beyond the capacity of someone without a law background. This is where your patent attorney should step in to offer professional legal opinion and advice. One has to note that most patent attorneys are paid by the hour, regardless of whether consultation is provided over the phone or in person. Hence, if the inventor has certain knowledge of IP law, the inventor can significantly reduce the number of hours spent with the patent attorney while ensuring that both the inventor and the patent attorney are on the same page so as to enhance efficiency in communication.

The inventor should furnish the patent attorney with the relevant and correct information with regard to his or her invention. Please be reminded that any lack of responsiveness could incur unnecessary legal cost and, as a result, reduce your business revenue. The inventor should collate and compile the complete details of all pertinent prior patents, publications or literature, and the results of any market research of similar products. A summary of all prior art is helpful. He or she should carefully prepare a document that encompasses important information, including the subject matter of the invention, a broad description of the invention, the problem statement and the advantages/features of the invention in providing a technical solution to the problem, the possible limitations of the invention, and any experimental or laboratory results.[23] Be sure to indicate in the document the inventor's name(s), company, and contact details, and the countries where patent protection is desired. To facilitate explanation of the invention to the patent attorney, the inventor should prepare some rough sketches, tables, graphs, or any other supporting materials to facilitate the understanding of the subject. It is also favorable if a glossary of terminologies or technical jargon that relates to the invention is prepared and submitted to the patent attorney.

While inventors have their own important role to play in the patenting process, the patent attorney practicing in the area of IC should also walk the extra mile to understand the IC invention as well as offer legal advice to the client. The patent attorney should keep up with the fast pace of technological advancements in the specific field of ICs. By doing this, patent attorneys will find themselves to be more appreciative of the inventions of their clients. This is a precondition essential in drafting a good patent document. On the other hand, once the inventor appoints his or her registered patent attorney(s), it also means that the inventor executes a power of attorney designating the named patent attorney(s) to act on his or her behalf. Following that, the patent offices conduct correspondence with the patent attorney(s) without involving or directly interacting with the inventor, although the latter is still allowed to communicate with the patent office regarding the status of his or her patent application.[24] Therefore, the patent attorney should also stay vigilant and responsive especially during patent examination or prosecution. This is important in forging a long-term and strong attorney–client relationship.

Summary

In this chapter, we explored the significance of IP rights for the IC industry. Nearing the turn of the 20th century, most advanced economies in the world evolved into knowledge-based economies (KBE), which depend profoundly on human cognition and inventiveness. The IC industry, immensely knowledge-based, has played an important role in sowing the seeds for the advent of KBE. In fact, it has been recorded that one of the first observations of knowledge being exploited as a power tool for economic growth emerged just a few years following the birth of ICs. Moreover, the IC industry is responsible for molding the present and future sustainability of KBE. In this context, intellectual property (IP) law, which is the legal underpinning of the IC industry, is an area worth pondering. Therefore, the onus is on IC designers, inventors, managers, and owners of companies who are involved in the heavily knowledge-based IC environment to prepare themselves with ample understanding of IPs and their underlying laws and principles.

Without proper IP support, IC innovations and scientific breakthroughs will most probably come to a grinding halt as inventors lose their momentum, because their rights are not protected, and they are not given fair incentives to move forward. In addition, IP protection creates IC business opportunities, drives new business models, and brings monetary rewards to individuals and enterprises. In fact, mere ideas or IPs without a business built around them are meaningless. IPs should, therefore, be accompanied by a strategic business methodology to capitalize on the idea. Businesses, on the other hand, should exploit and leverage IPs for revenue generation and for sustaining a competitive edge. With this in mind, we have provided extensive elaboration on the business applications of IP and its contribution to the economic progress for single inventors and for businesses or enterprises.

We have also devoted a separate section to describing IP reuse for IC design. IP reuse means reusing a previously designed and validated IP core or subcircuit in a new product. Efficient IP reuse often leads to a shorter time-to-market window, higher design productivity, cost effectiveness, less arduousness, and provides IC designers with more time to innovate. The promise of IP reuse in plugging the productivity gap has created an entirely new branch of the semiconductor industry that is based on the IP business model. Within the IP business ecosystem, three major players—IP provider, tool developer, and integrator—leverage on and bolster each other while promoting an atmosphere for innovation.

An investigation of the number of patents pertinent to ICs as issued by prominent patent offices in the world has also been analyzed in this chapter. The patent search was based on a five-year timeframe and carried out using a well-known research literature database. The patent analysis shows that there is an encouraging trend of ever-increasing patent counts since the early 1970s. We highlighted and explained the importance of establishing cooperative relationships between IC inventors and IP attorneys to achieve a win-win situation.

3

Technical Context of Integrated Circuits

The preceding chapter illustrated the importance of intellectual property (IP) for the field of integrated circuits (ICs) in considerable depth and breadth. We now provide a description of the technical context of ICs, because it is difficult for readers to appreciate and correlate IP for ICs without first understanding the IC from its most basic or elementary technical aspects.

At first glance, the mere mention or thought of ICs can be both intimidating and bewildering to many of us, especially for novice learners, lay persons, or attorneys without any background knowledge related to ICs. The field of ICs is indeed a vastly technical area of expertise that is constantly evolving, and it involves years of studies and work experience to master the intricacies. We acknowledge the fact that many nontechnical readers will shy away from this topic if we provide too much technical content of the IC. Therefore, it is our aspiration in this chapter to equip readers with only essential and fundamental knowledge of ICs and make the study of this chapter as basic as possible. We aim to provide a nontechnical guide to the technical framework of ICs, while ensuring that sufficient information is provided to paint the *big picture* of ICs. Armed with this knowledge of ICs, an IP attorney can communicate more effectively with an IC inventor, thereby bridging the technical gap between them.

Defining Integrated Circuit (IC)

> *What we didn't realize then was that the integrated circuit would reduce the cost of electronic functions by a factor of a million to one, nothing had ever done that for anything before.*
>
> *Jack Kilby, inventor of the IC*

The invention of the IC in the 1950s laid the conceptual and technical foundation for the entire field of modern microelectronics, propelling the industry to progress into a rapid technological advancement of miniaturization and integration that still continues to this day.[1] Now, you may ask: What is an IC?

An IC, also known as a chip, is defined as "a combination of interconnected circuit elements inseparably associated on or within a continuous substrate."[2] In layman's terms, an IC is a miniaturized electronic circuit comprising many transistors and other circuit elements that are designed and connected together to perform a certain function. These active and passive components are constructed on some kind of medium or substrate, usually silicon, and are meant to jointly function as a unit. The substrate is a platform or supporting material upon or within which an IC is manufactured. Active devices include transistors and diodes, whereas passive devices include, among others, capacitors and resistors: each with varying electrical behaviors, characteristics, and functionalities. In brief, diodes allow electricity or current to pass only in one direction and are often used for rectification purposes; capacitors collect electricity and are normally used to store energy; resistors restrict the flow of electricity and are able to control the amount of current flowing through a circuit path.

Before proceeding into more detail about ICs, let us take a step backward and look at something more basic: the transistor. The advent of the transistor in the 1940s provided a huge stimulus to engineers to design complex electronic circuits consisting of thousands of discrete components, including transistors, diodes, and capacitors which were unheard of during that period.[3] However, the potential technical showstopper at that time was not the large quantity of discrete components. Rather, the real difficulty lay in connecting up these components to form electronic circuits, because the soldering process involved was exceedingly time-consuming and expensive. In addition, due to the physical distances between the discrete devices, the critical speed bottleneck of electronic circuitries was soon hit. These technical challenges somehow motivated the engineering community to look for a more cost-effective and reliable way to produce these components and connect them, thus doing away with the painstaking, laborious soldering process. Consequently, the concept of IC was born.

Instead of having the discrete elements manufactured independently and then manually assembled on a printed circuit board (PCB) at a later stage, all the elements of an IC, including their interconnects, are fabricated at the same time on a single piece of substrate. This creates substantial advantages in terms of cost and performance. First, the fabrication cost of an IC is much lower than that of the discrete circuit because all the components of the chip and their interconnections are printed as a unit and not constructed one at a time. This leads to the additional advantage of the design being physically smaller in size. Secondly, the performance is greatly enhanced as the communication within circuits on-chip are much faster than between discrete components assembled on a PCB because the components are miniscule and in close proximity to each other. To summarize, the concept of ICs has brought a revolution to the circuit design arena and motivated the creation of highly sophisticated circuit structures that were uneconomical and close to impossible to build in the pre-IC era.

During a typical IC fabrication process, large batches of identical circuits are fabricated onto a single semiconductor wafer. As mentioned in Chapter 1, the wafer is the basic raw material used in the fabrication of chips and is made of semiconductor materials such as silicon. It appears as a thin slice of circular disc with a variety of sizes ranging from as small as 25.4 mm (1 inch) to the current industry standard of 300 mm (12 inches). Once the fabrication process is completed, the wafer is cut apart into smaller rectangular blocks, each having a replica of the circuit. Each of these blocks is called a die, and is individually packaged (or encapsulated) after being tested. A pictorial illustration of a wafer and the dies is presented in Figure 3-1. The packaged semiconductor die is also called an IC chip. ICs are

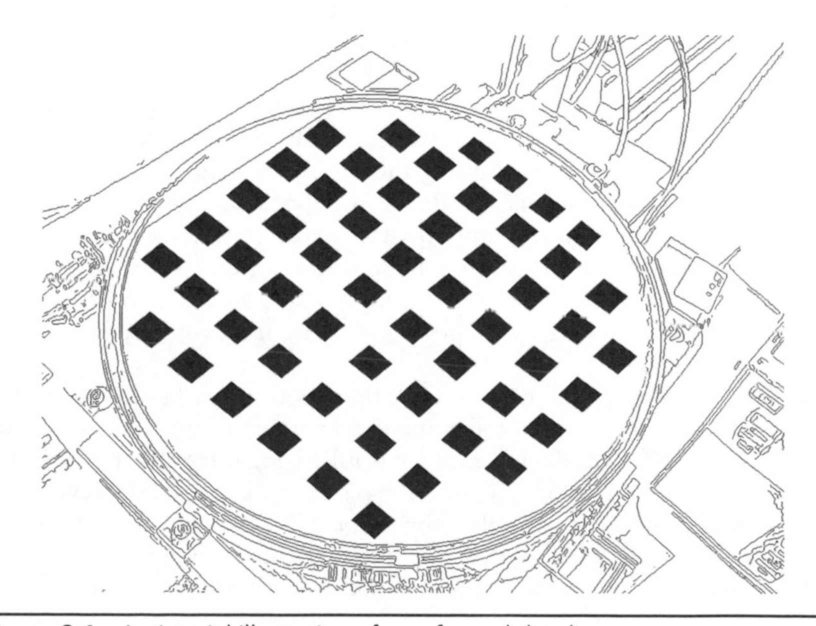

Figure 3-1 A pictorial illustration of a wafer and the dies

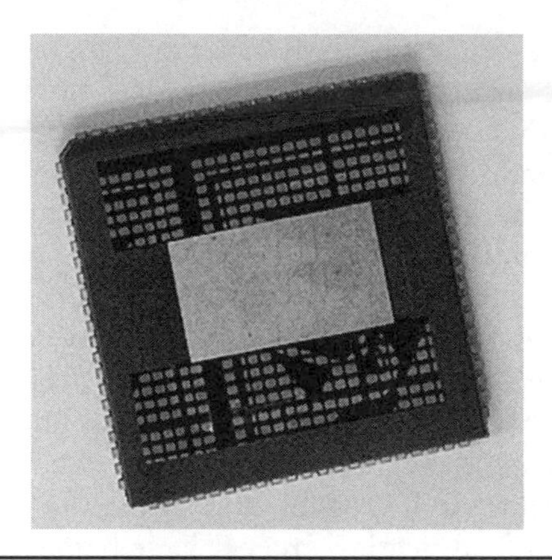

Figure 3-2 An example of an IC chip

known as bare dies, or simply dies, until they are packaged. In practice, the terms IC, chip, and die are normally treated as synonyms and are frequently used interchangeably. Figure 3-2 shows an example of an IC chip.

Evolution of IC Technology

At present, there is a broad variety of IC technology available for electronic designers. This ranges from the older general-purpose technologies such as the bipolar category of Transistor-Transistor Logic (TTL or T^2L) and Emitter-Coupled Logic (ECL), the high-density and low-power technology of the Metal-Oxide Semiconductor (MOS) process, to other more application-specific technologies such as Gallium Arsenide (GaAs) and Silicon Germanium (SiGe). Among these technologies, MOS is the most prevalent technology and has escalated in importance over the past three decades, while providing the impetus for a host of state-of-the-art digital Very-Large Scale Integration (VLSI) applications. Bipolar, GaAs, and SiGe circuits are also in use, albeit representing a smaller fraction of the total consumer market. In particular, these circuits are used when speed is of the utmost importance.[4]

In fact, as we trace the historic trail, the bipolar logic families have seen their golden days in the past, following the invention of the bipolar junction transistor in 1949. The first truly successful IC logic family, the TTL, was based on bipolar technology.[5] The TTL has since kick-started the revolutionary age of ICs, and it was also the fabrication of TTL devices that gave birth to the inauguration of large-scale semiconductor pioneering companies such as Fairchild and Texas Instruments.[6] Meanwhile, the ECL logic family was deemed to yield greater advantage when high-speed operation is required. The bipolar technology continued to flourish and dominated the digital semiconductor market until the 1980s when its use started to decline.

Even though bipolar technology renders high-performance features and good noise performance, it is inherently plagued with extremely high-power dissipation—a stumbling block that has impeded its prolonged usage beyond the 1980s. This drawback places a stringent restriction on the number of transistors that can be integrated on a die without any reliability problems, hence limiting the IC integration density. This issue, which remains unresolved, has caused bipolar logic to assume defeat in the battle of hegemony in the digital design arena. IC technology development, as a consequence, has arrived at a crossroads. Ultimately, a dramatic technological switch has been made toward MOS technology due to its high density of integration.

Through the 1970s, the pure *n*-channel (nMOS) process was commonplace and coexisted with the TTL logic from the bipolar category. The nMOS retained its importance until the late 1970s when it began to experience the same technical crisis that made high-density bipolar logic unattractive or impractical, namely the power consumption constraint.[6] Henceforth, a gradual transition to the Complementary Metal Oxide Semiconductor (CMOS) technology took place. Since then, CMOS technology has become the prevailing fabrication technology for memories and microcomputers due to its low-power and high packing density characteristics. This remains true. In the mid-1980s, a better speed performance than that achievable by CMOS was required, and BiCMOS (Bipolar Compatible CMOS) technology was seen as an alternative to CMOS in some applications. Therefore, while CMOS maintains its position as the most dominant process on the present IC manufacturing scene, other technological advances such as the BiCMOS, GaAs, or SiGe approaches coexist, but they are used only for specific applications that require exceptionally high performance.

As briefly mentioned above, the performance and power inadequacies encountered in bipolar and CMOS technologies have given birth to BiCMOS technology, which is essentially the marriage of *the best of both worlds* by merging bipolar and CMOS devices on the same substrate.[7,8] The BiCMOS technology exploits the specific advantages of the properties of the two originally stand-alone technologies by appropriately trading off characteristics of each technology; while the bipolar technology offers high current-driving capability, high-switching, and input/output (I/O) speed, and good noise performance, CMOS technology promises low power dissipation, high packing density, and high noise margin. Thus, BiCMOS technology permits performance optimization and a higher degree of system integration with an improved speed over CMOS and lower power dissipation than that of bipolar technology. Hence, an ultimate balanced performance combined with the power characteristic is achieved.

However, the realization of these advantages does not come free. Instead, it comes at the expense of process complexity that involves complex technology development, intricate chip production tasks, and much higher manufacturing costs. Furthermore, this greater process complexity incurs longer fabrication cycle time as compared to either bipolar or CMOS technology. BiCMOS technology continued to enjoy popularity until another important development took place in the early 1990s. During that period, CMOS designers began to scale down power-supply voltage in earnest[9–11] and, consequently, the performance gap between bipolar and CMOS for digital logic began to narrow. Pursuant to this technological evolvement, the initial rationale and the expected advantage of enhancing the chip or system performance by integrating bipolar transistors on a CMOS chip gradually diminished. Therefore, BiCMOS technology was not pursued for digital applications, especially those that could settle at low or medium performance. Today, CMOS technology has carved a niche for itself as the feasible semiconductor VLSI technology for microprocessors, memories, and application specific integrated circuits (ASICs). Frequently, CMOS turns out to be the top choice for digital-logic design, particularly in portable, low-cost, and low-frequency applications.

Applications of ICs

ICs can be categorized with respect to their applications—digital, analog, and mixed-signal.[12] The major difference between the various applications of ICs is in the way quantities are represented in their operation. As a matter of fact, in our daily lives, we are constantly dealing with quantities. Quantities exist in science, technology, commerce, and almost all other facets of human endeavor. Therefore, there should be a way, or ways, to deal with these quantities, and it is always crucial to convey the values reliably and efficiently without compromising its accuracy. The typical ways of representing numerical values of quantities in ICs are analog and digital. Analog ICs involve quantities with continuous values, whereas digital ICs involve quantities with discrete sets of values. There are also ICs that involve both digital and analog representation of quantities on a single chip. These hybrid ICs are appropriately named mixed-signal ICs.

Analog ICs

The real world we are living in is fundamentally analog in its innermost nature. Most phenomena or things that can be measured in physical quantities appear intrinsically in analog form: sound, light, distance, temperature, you name it. Therefore, it is this analog representation of quantities that is often taken as the inputs and outputs of a device or system that receives, monitors, processes, and controls the quantities. These analog devices monitor real world events that vary over a range of values in a continuous fashion, such as movement, sound, and temperature, and convert them into analogous electronic signals or mechanical patterns that are closely proportional to the actual signals. An example is the analog clock that manifests the rotation of the planet earth with the continuous motion of rotating pointers or hands around the clock face.[13] At any given moment, the planet is rotating. Hence, time as indicated on an analog clock does not go from, say 11:30 a.m. to 11:31 a.m. instantaneously, but rather it transitions through all the infinite values in between. Similarly, the temperature in the atmosphere during autumn at a certain location does not change abruptly from 18° C to 19° C, rather, it fluctuates with all possible intermediate values in between. A graphical representation of the temperature portrays a smooth and continuous waveform such as shown in Figure 3-3.

In analog electronic systems, a continuously varying signal is normally represented by changes in voltages, currents, or frequencies that are proportional to the values of the signal. For instance, a telephone picks up voice vibrations and converts them into electrical signals of identical shape before transmitting them via telephone lines. Analog ICs include operational amplifiers, oscillators, comparators, and sensors that carry out operations such as amplification, modulation and demodulation, and filtering. Moreover, the escalating importance of the wireless industry has also spurred the advent of an important segment of the analog IC. In fact, wireless technology and telecommunication have been evolving dramatically with enhanced and novel applications emerging almost every other day.[14] These applications necessitate high-frequency carrier signals to ensure efficient transmission

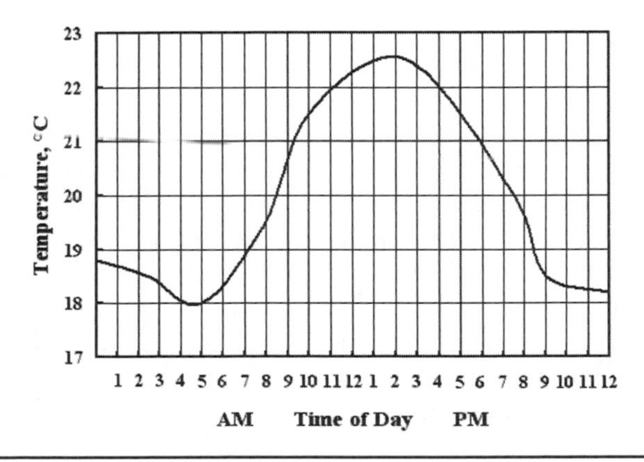

Figure 3-3 An analog waveform

of information, and the design of these high-frequency analog circuits, radio frequency ICs, or RFIC, differs from conventional low-frequency analog circuits. The four main markets for RFIC, typically in the frequency range of 0.8 MHz to 3 GHz, include wireless phones (cellular communications and cordless phones), networks (wireless local area networks, [WLAN]), positioning (global positioning systems, [GPS]), and sensors (material detections such as keyless door locks).[15] Automatic identification of remote data using radio-frequency identification (RFID) tags, cordless computer mice and keyboards, and satellite television broadcasts are some other areas of applications of RFIC.

The unique ability of the analog approach to capture the subtle nature of real world events is perhaps its only benefit when compared to its digital counterpart.[13] Ironically, upon hindsight, this advantage of analog technique also infers its vulnerability to noise-induced disturbances. Analog devices capture every possible detail of an event, and any small change or random disturbance is analogous to a distortion in the original signal—signal degradation may occur, and in some serious cases, the information may be lost altogether. In reality, nearly all tasks that can be performed by analog ICs can also be executed in the digital realm. Digital ICs have been soaring in importance and have penetrated throughout the entire electronics market.

Digital ICs

The present technological era is commonly known as the digital age—this suggests the imperative role of digital systems in our everyday life. Digital information, which is basically just a bunch of numbers, is easier to store and manipulate when compared with its analog counterpart. Additionally, digital technology provides greater data compression capability hence allowing higher IC integration density and more functionality, as driven by the incessant scaling down of process technology. Digital ICs, such as microprocessors used in computers, digital signal processors, and microcontrollers, can contain billions of transistors within a die of a few square millimeters. The minute size of these devices renders design merits such as low-power consumption, high operating speed, and reduced manufacturing cost when compared to discrete circuits or analog devices.

While analog ICs deal with continuous signals, digital ICs manipulate discrete (noncontinuous) signals or discrete units of information.[16] Any set of data that is confined within a finite number of units comprises discrete information: the six strings of a classical guitar, the 10 decimal digits, the five-speed manual transmission of a vehicle's gearbox, and the 26 letters of the alphabet are such examples. Many number systems can be used to convey digital or discrete information, the most common methods being decimal, binary, octal, and hexadecimal. Among these, the most popular form of number system that is in use in modern digital applications today is the binary representation. A binary digit, also known as a bit for short, has only two possible states: 0 (zero) and 1 (one). When they are grouped together in a series of 0s and 1s, called binary codes, they can be used to convey digital information or any quantity represented by the other number systems. For example, the decimal number 9 can be correspondingly represented by a 4-bit binary code of 1001. In digital ICs, each bit is represented by electrical signals, typically a voltage or a current. Only discrete levels of a particular signal are considered, with no intermediate values, which is what makes the digital approach less susceptible to noise compared to the analog technique. A signal level of high (usually the power supply voltage) or low (usually the ground potential) denotes a logic level 0 or 1 of the binary value. Modern digital IC design generally adopts the active-high logic convention, whereby a high voltage (or current) level represents a binary 1, and a low voltage (or current) level indicates a binary 0. The active-low logic, which is the reverse of its active-high representation, is only used provided it is explicitly mentioned. A graphical representation of a typical digital waveform is shown in Figure 3-4, whereby the logic levels are denoted in voltages.

Other than inherently discrete sets of information, such as those cited in the previous paragraph, digital quantities are often derived from analog signals. In other words, digital representation is a sampled approximation of an analog phenomenon. Digital ICs work by chopping and formatting the originally continuous signal into binary codes in which the information is represented by a series of 1s and 0s. Let us refer back to our clock example, but this time a digital one, for a good comparison with the analog version. A digital clock is only capable of representing a finite number of times. It shows the time in decimal digits of hours and minutes (and in some cases seconds or even tenth of seconds). Recall that our earth rotates relentlessly thus time also moves on continuously. However, if you look at your digital clock on the wall, its reading does not change constantly; instead, it renews its reading at some intervals, i.e., in steps of one per minute or second. This implies that the digital portrayal of the time changes in a discrete or step-by-step manner, which is in contrast to an analog clock, whereby the change in time reading is continuous.

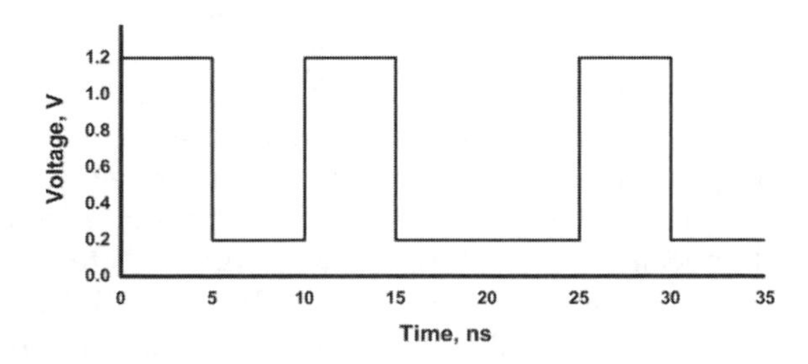

Figure 3-4 A digital waveform

Mixed-Signal ICs

With the huge market demand and incessant quest for portability and functionality, there is a growing trend of complex integrated systems such as system-on-chip (SoC) emerging in the mass consumer market. The move to SoC is further fueled by the continuous downscaling of transistor feature sizes that increasingly allows an entire system to be designed into a single integrated device, while garnering the advantages of cost, space, and power reductions, as well as enhanced chip performance.[17] These leading-edge integrated systems often encompass hybrid or mixed-technology designs in which high-performance analog or mixed-signal blocks, and perhaps sensitive RF frontends, coexist with complex digital components.[18] Put simply, mixed-signal ICs consist of both analog and digital circuits on a single die.[19]

A mixed-signal IC, for example an analog-to-digital converter (ADC), enables a system to accept analog information from the real world as its input and convert it into binary form.[20] This binary representation of the signal is then processed arithmetically within a digital system. Analog-to-digital conversion takes place when your favorite song album is recorded on an audio compact disc (CD) at a music studio, where the analog sound waves are converted into digital information that will be stored on the disc. Whenever you feel like listening to the songs recorded on the disc, the digital data is first being read and then transformed back into an analog signal, which is an electrical reproduction of the original sound by using a digital-to-analog converter (DAC). Subsequently, this analog signal is amplified and transmitted to the speaker for your listening pleasure—this is the fundamental operating mechanism of a CD player.

Semiconductor and *p-n* Junction

Semiconductor actually consists of two words: *semi* and *conductor.* The first word *semi* means intermediate whereas the second word *conductor* means a material that allows electricity to flow easily. It is also useful to note that the antonym of conductor is insulator which prohibits the flow of electricity. Hence, taken as a word, semiconductor refers to materials whose conductive properties (the ease with which electricity can flow through the material) lie between that of the conductor and insulator. The conductivity of a material within a certain distance is measured in Siemens/cm, or simply S/cm. For example, the conductivity of copper, a good conductor, is 0.59×10^6 S/cm whereas the conductivity of glass, an efficient insulator, may range between 10^{-16} and 10^{-13} S/cm. As expected, silicon, a semiconductor, has a conductivity of between 10^{-8} to 10^{-1} S/cm which lies between those of copper and glass.[21] Nonetheless, in our day-to-day usage, the word *semiconductor* is normally used to refer to electronic devices fabricated from semiconductor materials, most frequently ICs.

The basic semiconductor wafer, which is fabricated from pure silicon, is by itself not useful from the viewpoint of IC technology. Even though the conductivity of silicon lies between that of a conductor and insulator, it is still too low to be of use for circuit designers. Hence, it is of the utmost importance that the conductivity of certain areas of the semiconductor wafer be tailored to meet its specific objectives. For example, areas in the semiconductor wafer whereby the IC is fabricated should have conductivity superior to that of pure silicon while areas in which no IC is fabricated should retain the conductivity of silicon. This process of tailoring the conductivity of selected areas in the semiconductor wafer is called doping and is part of the fabrication process.

Doping is a process in which impurities are added to the crystalline structure of the silicon. These impurities can either be donor or acceptor elements depending on the type of carriers they introduce into the undoped silicon. The former (donor) introduces an abundance of free electrons, and, hence, the doped silicon is called an *n*-type semiconductor whereas the latter (acceptor) introduces holes as the majority carrier (in effect positive charges) and the resultant doped silicon is called a *p*-type semiconductor. Consequently, the *n*-type semiconductor (electron rich) conducts current predominantly by free electrons whereas the *p*-type semiconductor's main conducting agent is the holes (hole rich). The reader will discover that both *n*-type and *p*-type semiconductors form an integral part of the transistors in the next section.

Types of Transistors

A transistor is a semiconductor element having typically three to four terminals (electrical contacts). It is capable of performing two basic functions that are critical to the design of electronic circuits—amplification and switching. In simple terms, amplification refers to the process of magnifying the strength of an electronic signal by means of energy transfer from an external source. Meanwhile, switching is the process of manipulating a current flowing through or voltage across two terminals by using a small current or voltage applied at a third terminal.[21] Put simply, the fundamental function of a switch is to turn the electricity on or off. The types of transistors available on the IC market are determined by the IC fabrication technologies used to produce them. The two families of transistors used in mainstream ICs are the metal-oxide semiconductor field-effect transistors (MOSFET) produced in a MOS or CMOS environment and the bipolar junction transistors (BJT) produced by bipolar technology—these two types of transistors are our primary focus in this section. The discussion of MOS transistors precedes that of the BJT because it is the dominant IC technology in the present day. However, please take note that this is by no means a complete coverage of the said transistors. Our objective is to prepare the reader with the basic understanding of the devices at the elementary level. Readers who are interested in a more detailed and comprehensive description of transistors are encouraged to refer to books or literature specializing on the subject.

Metal-Oxide-Semiconductor Field-Effect Transistor

The metal-oxide-semiconductor field-effect transistor (MOSFET) is so called because the first MOS circuits realized in the early 1970s were fabricated based on metal-gate *p*-channel (pMOS) technology.[22] However, since the 1980s, polysilicon (as its name implies, it is a type of silicon) has replaced metal as the dominant gate material due to the former's ease of deposition in the manufacturing process and its excellent tolerance to excessively high temperature (⊠ 900 °C) encountered in subsequent fabrication steps. Nonetheless, the name MOSFET still remains in use in the IC industry notwithstanding this replacement.

The MOSFET transistor is a four-terminal device constructed by doped regions of semiconductor material. Each region is connected to a terminal and is appropriately named gate (G), source (S), drain (D), and bulk or body (B). A voltage applied to the gate determines the on- or off-state of the transistor. There are basically two types of MOSFET devices—the nMOS transistor and the pMOS transistor. The nMOS transistor is constructed with n^+ drain and source regions embedded in a *p*-type substrate, whereas the pMOS transistor consists of p^+ drain and source regions embedded in an *n*-type substrate.[6] The superscript of a plus or (+) sign relatively represents a heavily doped material. For example, the n^+ drain means that the semiconductor material is doped heavily with the donor elements while the p^+ drain means that it is heavily doped with the acceptor elements.

The three-dimensional (3-D) perspective views of the nMOS and pMOS physical structures are shown in Figure 3-5 and Figure 3-6, respectively. To establish a better understanding for the reader, we have segmented the 3-D view of each type of the MOS device into their top views and cross-sectional views, and these are correspondingly presented in Figures 3-7 and 3-8. It is worth mentioning here that the top view representation of the transistor is actually what an IC designer usually *sees* while drawing the layout of the circuit in a computer-aided design (CAD)* environment. In other words, the top view representation is the two-dimensional perspective of

---

*A more detailed discussion of CAD is provided later in this chapter

the three-dimensional disposition. It is also important to bear in mind that the drawings illustrated in Figures 3-5 through 3-8 are the most simplified versions of the layout-designs—an actual IC consists of many more layers of metals, insulating and semiconductor materials, which are stacked on each other.

In circuit design, the fourth terminal, the body or bulk (both used interchangeably with the substrate), is normally treated as secondary to other terminals. It is by default assumed to be tied to a dc supply, i.e., the power

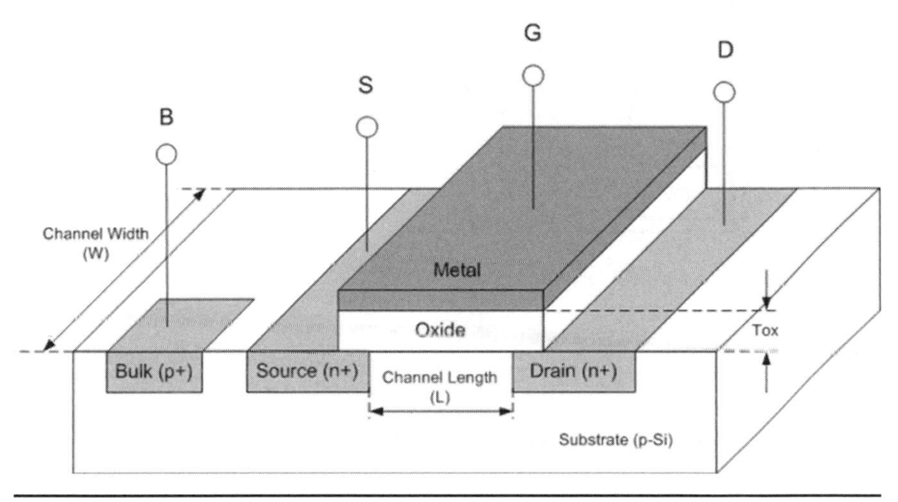

Figure 3-5 Physical structure of an nMOS transistor

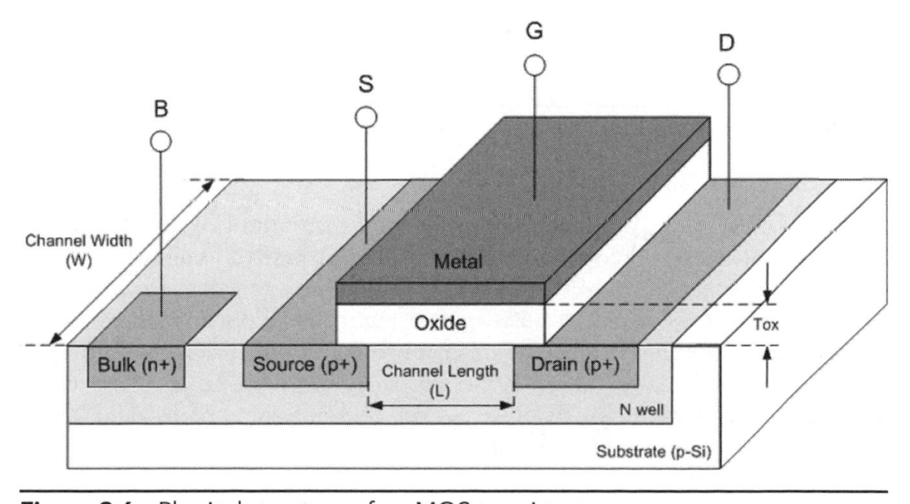

Figure 3-6 Physical structure of a pMOS transistor

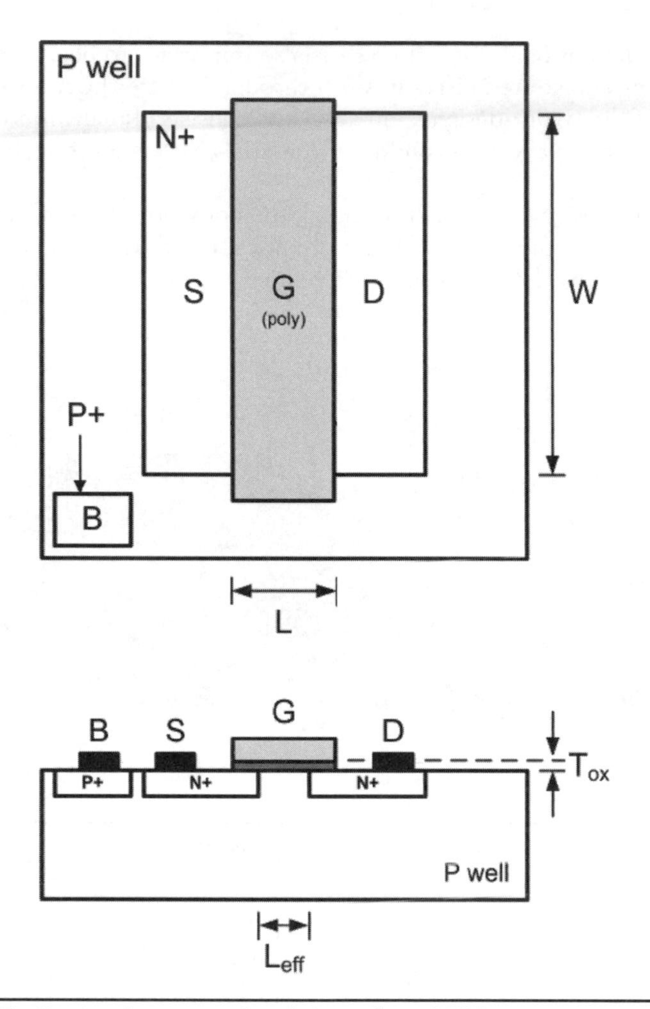

Figure 3-7 Top- and cross-sectional view of an nMOS transistor

supply voltage (V_{DD}) for pMOS and ground potential (GND) for nMOS if it is not shown in the symbol, in which it is presented as simply a three-terminal device. Figure 3-9 shows the four-terminal and three-terminal circuit symbols of the nMOS transistor and Figure 3-10 portrays those of the pMOS transistor. Despite the fact that pure nMOS or pMOS technologies are seldom used today, these MOS transistors form the basis of the vastly popular complementary MOS (CMOS) technology, which yields both nMOS and pMOS devices on one chip.

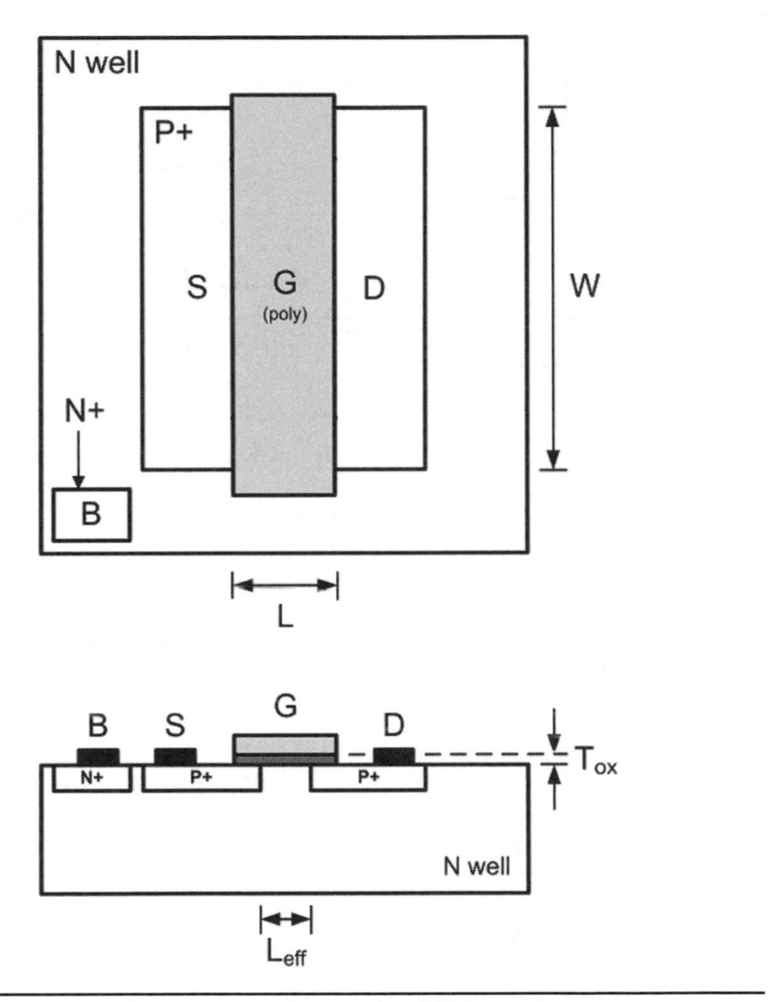

Figure 3-8 Top- and cross-sectional view of a pMOS transistor

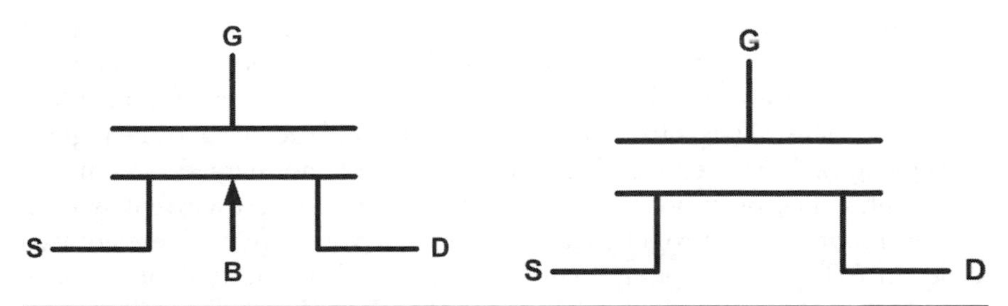

Figure 3-9 Circuit symbols of nMOS transistors (a) four-terminal device. (b) three-terminal device

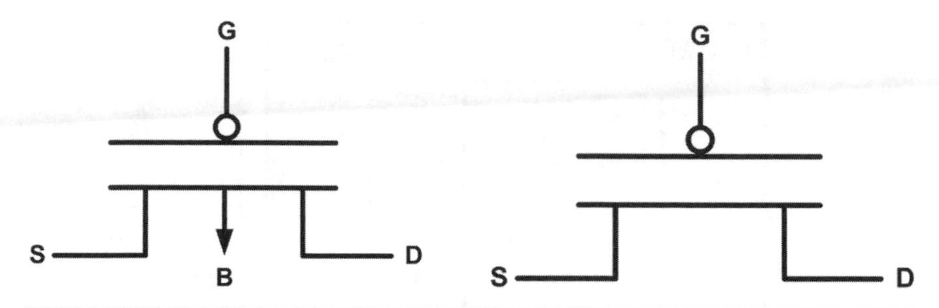

Figure 3-10 Circuit symbols of pMOS transistors (a) four-terminal device. (b) three-terminal device

One good example of a CMOS circuit is an *inverter* whose function is to invert the incoming signal. Figure 3-11 shows the schematic diagram alongside the layout of a typical CMOS inverter. It consists of only two transistors, an nMOS and a pMOS, which are connected in a complementary way to form a logic gate. Here, the gates of both MOSFETs are connected together and form the input of the inverter. Also, both the drains of the nMOS and pMOS devices are directly connected to the output and the source of the nMOSFET and pMOSFET are connected to GND and the V_{DD}, respectively. The CMOS inverter operates in this manner: If the input logic voltage is a 1 (V_{DD}), then the nMOS will turn on because it has a positive gate to source voltage (V_{GS}) whereas the pMOS will be off. This phenomenon provides a path from the output (V_{out}) to GND. The output terminal is essentially shorted to ground and isolated from the positive terminal of the power supply. The output level is therefore a '0'. On the other hand, if the input level is a 0, the reverse is true. The pMOS transistor will conduct, and it provides a path from V_{DD} to the output (V_{out}). The nMOS transistor is isolated, and the output level is a 1.

Bipolar Junction Transistor

The name bipolar junction transistor (BJT) was coined due to its operation involving the conduction by two types of carriers—electrons (negatively charged) and holes (positively charged), both within the same semiconductor crystal. The BJT is a three-terminal device constructed by three regions of separately doped semiconductor material, explicitly named the emitter, base, and collector regions. In most cases, the small base current is used to control or yield a proportionally larger collector current. There are two types of BJT devices, *npn* and *pnp,* as determined by the type of majority mobile carriers or, simply, the type of material used in forming the emitter, base and collector regions. In an *npn* transistor, the three regions are

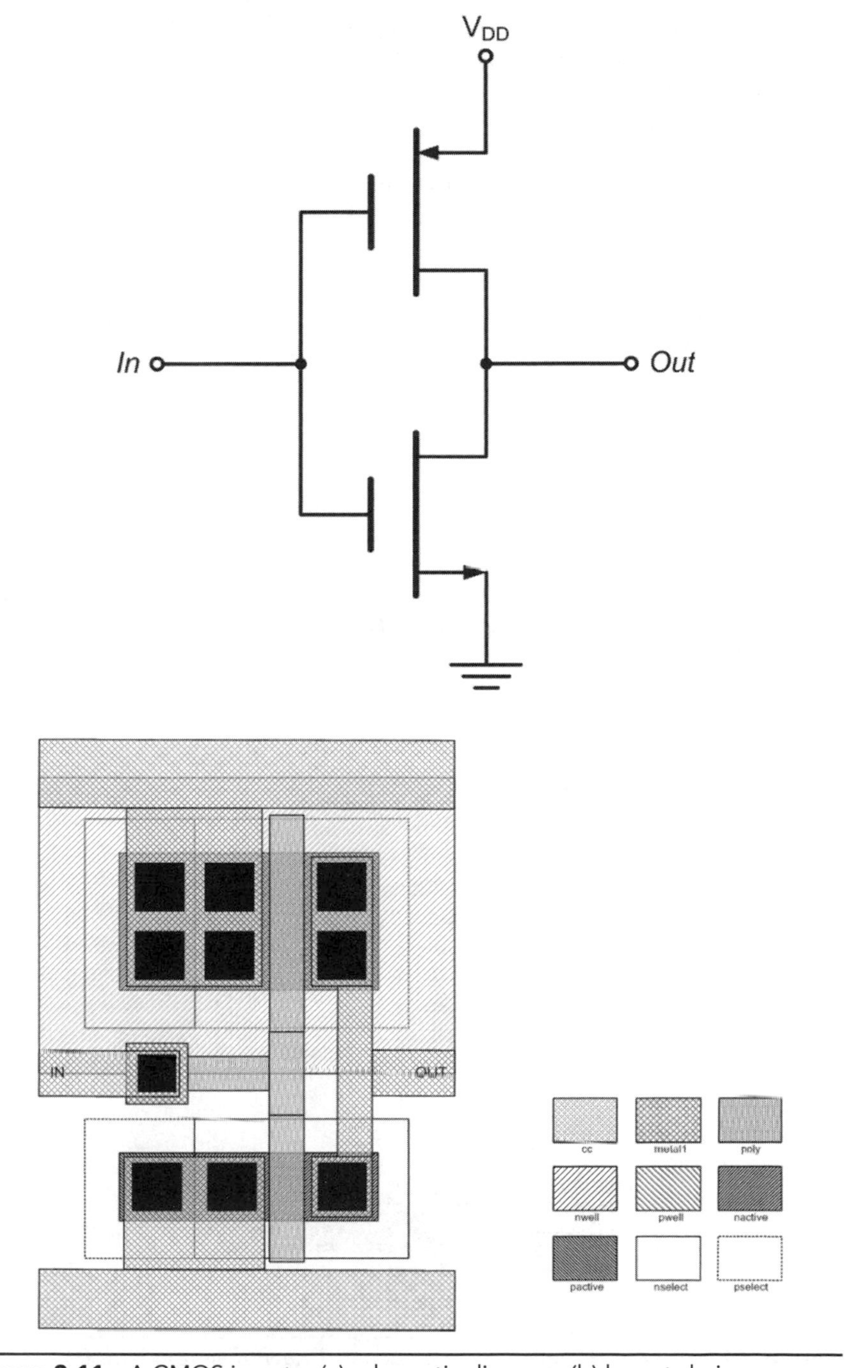

Figure 3-11 A CMOS inverter (a) schematic diagram. (b) layout-design

respectively of the *n*-type, *p*-type, and *n*-type. These regions are independently tied to a terminal or contact, appropriately labeled emitter (E), base (B), and collector (C). This is illustrated in Figure 3-12, in which the circuit symbol of an *npn*-type BJT is portrayed. Figure 3-13 shows the top-view (i.e., the layout drawing) and the cross-sectional view of the *npn* device. The reader can actually visualize a three-dimensional perspective view of the *npn* physical structure by combining both top- and cross-sectional views. Similarly, a *pnp* transistor is so named because of its emitter, base, and collector regions, which are made of *p*-type, *n*-type, and *p*-type semiconductor materials. Figure 3-14 shows the circuit symbol, whereas Figure 3-15 depicts the layout drawing and cross-section of a *pnp* transistor.

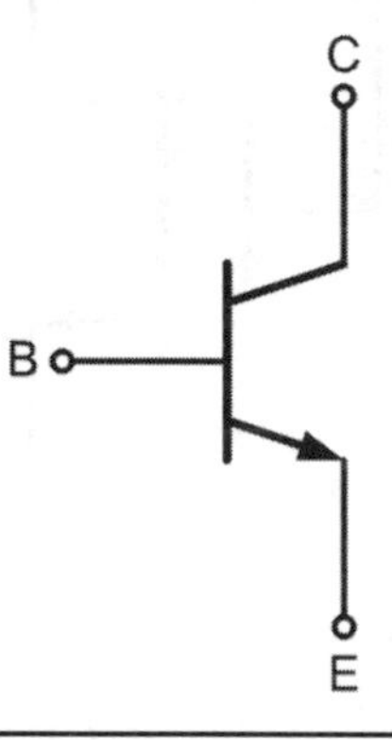

Figure 3-12 Circuit symbol of an *npn*-type BJT

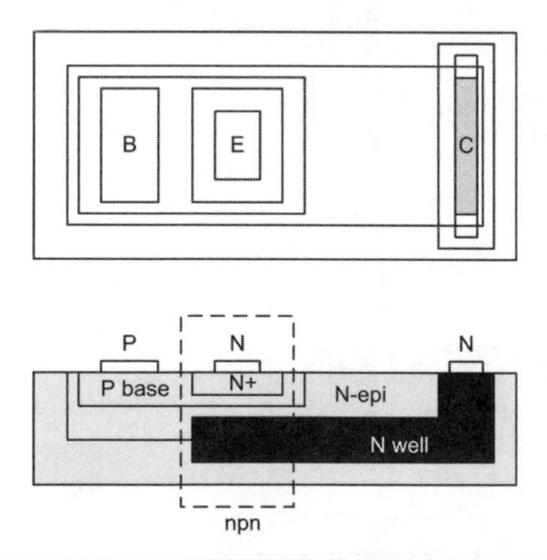

Figure 3-13 Layout-design and cross-sectional view of an *npn*-type BJT

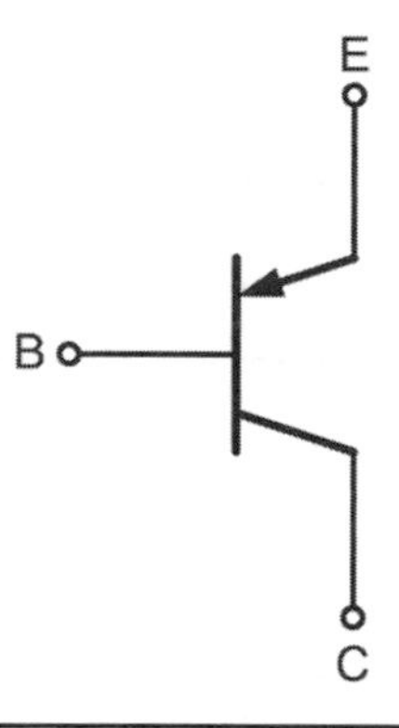

Figure 3-14 Circuit symbol of an *pnp*-type BJT

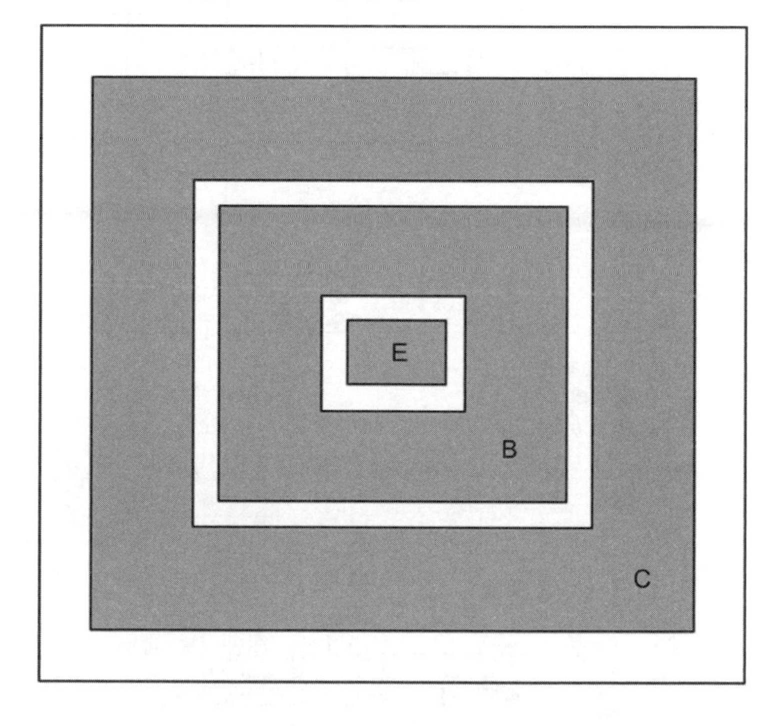

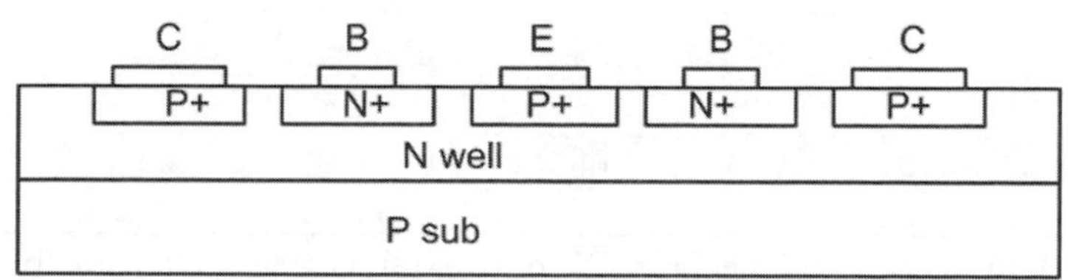

Figure 3-15 Layout-design and cross-sectional view of an *pnp*-type BJT

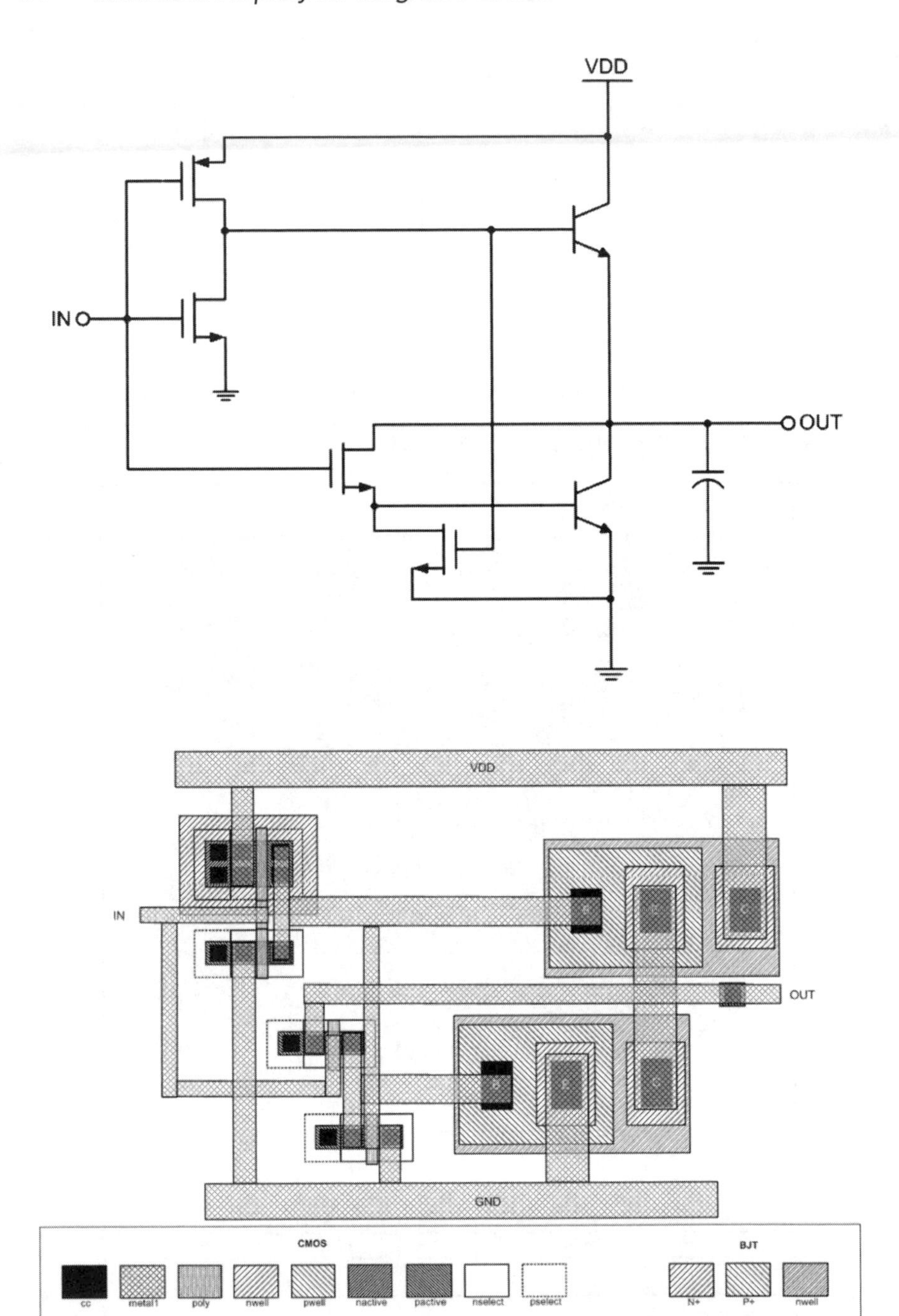

Figure 3-16 A conventional BiCMOS inverter (a) schematic diagram. (b) layout drawing

Bipolar Compatible CMOS

As its name implies, the bipolar compatible CMOS (BiCMOS) technology is a combination of bipolar and CMOS technologies. As such, this category of IC technology is built based on the two types of transistors we have discussed in previous sections, BJT and MOSFET. Most BiCMOS chips tend to be CMOS-intensive due to the stringent restriction in power consumption in high-density VLSI systems. Moreover, CMOS transistors are employed as the dominant devices for logic operations since bipolar transistors are mainly used only when high drive capability is required.[4] Figure 3-16 shows a conventional BiCMOS inverter and its corresponding layout drawing. This is a good circuit configuration to show how bipolar and CMOS technologies are integrated, while retaining the features of each individual device. The addition of the bipolar driver stage in a totem-pole configuration to the basic CMOS stage is responsible for the high current driving capability of the BiCMOS inverter over the CMOS counterpart, especially when the load capacitance is large.

Technology Node and Level of Integration

Significant progress has been made toward producing devices with ever-decreasing miniaturized scale. In this context, technology node is a term used to denote the minimum geometry or feature size that can be realized on a chip particularly in the field of digital ICs. It normally refers to the minimum dimension of the polysilicon used to form the gate of a MOS transistor. Hence, the more advanced the technology node is, the smaller the transistor will be. For example, a 0.18 μm (unit pronounced as micron*) technology node will provide a smaller transistor when compared to its 0.25 μm counterpart. Thus, one thing leads to another; smaller transistors lead to a higher level of integration (in other words, higher integration density) and faster switching speed, which in turn leads to increasingly high-performance electronic systems with numerous applications in computing, telecommunication, and signal processing.

Over the years, the technology node has been scaled down at a somewhat astounding pace, which is essentially a good testimony of the rapid advances in IC fabrication technology. This evolvement has further pushed the envelope of IC integration density. The IC industry started in the late 1960s and early 1970s with a 10 μm technology. In the mid-1970s, the popular technology node used was 7 μm. In the late 1970s and early 1980s, feature size shrank to 5 μm and further scaled down to 1 μm in 1989. In

*Short for micrometer, micron is one millionth of a meter; 1 μm = 1×10^{-6} m

a matter of years, the industry evolved into deep submicron CMOS technologies with a minimum feature size of 0.25 μm realized by 1995 and in 2001, the 0.18 μm was already commonplace.[4] It has been anticipated that nanometer-scale devices will continue this trend of device miniaturization. Today, that projection has become a reality and the IC industry has been greeted with technologies of the nanometer* range. In January 2008, Intel rolled out processors built on 45nm technology.[23] When this manuscript was prepared, some leaders of the IC industry were already venturing into the 32nm realm and beyond.

Contemporary ICs are packed with virtually limitless functions. They are fabricated with large-scale batch processes with unprecedented levels of integration and sophistication. The first monolithic† integrated circuit, which was fabricated in 1958, had only one transistor and after barely four decades, by 2001, the Pentium IV processor by silicon giant Intel comprised 42 million transistors.[24] In February 2008, Intel launched the first-ever IC chip that consisted of more than two billion transistors.[25] The evolutionary growth in the number of transistors capable of being contained on a single chip was made possible by the incessant miniaturization of transistor feature sizes (see the advancement in the technology node discussed in the preceding paragraph), and its evolution has been more drastic than expected. As such, ICs are often categorized according to the complexity of the circuits, as measured by the number of transistors or logic gates that are embedded within them. As a rough estimate, a logic gate typically contains 10 to 100 transistors.

Shortly after the advent of ICs, the process that produced the generation of ICs of less than 10 logic gates was known as small-scale integration (SSI).[16] As IC designers began to dip their toes into circuits with higher complexity, medium-scale integration (MSI), which was applied to ICs with approximately 10 to 1000 gates, was introduced in the mid-1960s. As the technology continued to advance in earnest, a single chip could achieve an integration of up to thousands or tens of thousands of gates and was known as large-scale integration (LSI). In the 1980s, the very-large scale integration (VLSI) technique surfaced and could produce hundreds of thousands of gates on just one chip. Following that, the ultra-large scale integration (ULSI) era allowed a single chip to easily integrate more than one million gates. Today, the collective terms of SSI, MSI, LSI, VLSI, and ULSI are customarily used to represent the number of logic gates packed on an IC chip according to their respective complexity.

---

*Nanometer (nm) is one billionth of a meter; $1 \text{ nm} = 1 \times 10^{-9}$ m

†A monolithic IC is defined as "an IC whose elements are formed in situ upon or within a semiconductor substrate with at least one of the elements formed within the substrate"[2]

Design Automation

Due to the never-ending demand for an ever-increasing level of sophistication of contemporary electronic systems coupled with the tightening time-to-market constraint and product obsolescence pressure, a handcrafted-based design approach for the design and analysis of ICs is almost impossible. This is where the thriving market of electronic design automation (EDA) or computer-aided design (CAD) design toolsets (the two terms are often used interchangeably) comes into play. The relationship between IC design and EDA is of a *positive feedback loop* nature. This means that while the incessant demand for higher circuit design complexity and shorter time-to-market are the primary factors that motivated the emergence of EDA tools and forced EDA vendors to improve the EDA tools significantly, on the other hand, the significant advancement of the EDA tools is the enabling factor that has propelled circuit complexity to move to unprecedented levels and made shorter time-to-market possible.

EDA/CAD tools, which permit the automation of a specific design step, generally appear in the form of a software program. An extensive variety of CAD tools are available on the commercial market, which include simulations at various levels of circuit complexity and design step: circuit, logic, layout, or architecture; design synthesis and analysis, design verification, and layout generation.[6] For instance, CAD systems allow circuit designers to simulate a two-dimensional circuit, commonly known as a circuit schematic diagram, by using standard electronic symbols to represent various circuit elements such as transistors, resistors, or capacitors on the computer. The CAD system also simulates circuit operation, evaluates performance, verifies transistor and circuit parameters, and detects possible temperature, voltage, speed, or timing problems. The circuit/device and design parameters, such as transistor sizes, temperature, and operating frequency can also be tweaked, swept, and optimized to realize the maximum potential of the circuit. Some examples of these tools and the EDA vendors that provide them are the Virtuoso platform (normally used for analog, mixed-signal, and RF designs) and the Encounter platform (for designing digital ICs) from Cadence Design Systems, PrimeTime and Design Compiler from Synopsys, and ModelSim from Mentor Graphics.

The present design automation technology for analog and mixed-signal IC is lagging several generations behind existing digital technology.[26] This is mainly attributed to the nature of analog and mixed-signal designs, which are essentially heterogeneous, more intricate, more subtle, and less hierarchical in comparison to digital design. The lack of an efficient CAD tool for the analog part of the chip poses a design bottleneck on the overall productivity of contemporary SoCs.[19] In this respect, ideas and design methodologies, such as the realization of CAD-supported hierarchical design flow that facilitates design reuse or IP reusable blocks, have been actively researched and proposed in the literature.[26,27]

IC Design Flow

Depending on its complexity, the design process of an IC chip can be costly, arduous, and lengthy. It can take several months or even multiple years inching toward its successful completion. In general, IC design requires numerous iterative steps before the final chip design gradually takes shape.[28] The flow chart of a typical full-custom IC design procedure is presented in Figure 3-17. First and foremost, design specifications have to be derived and settled upon based on specific engineering requirements and system criteria, which is generally the result of the translation of business objectives.

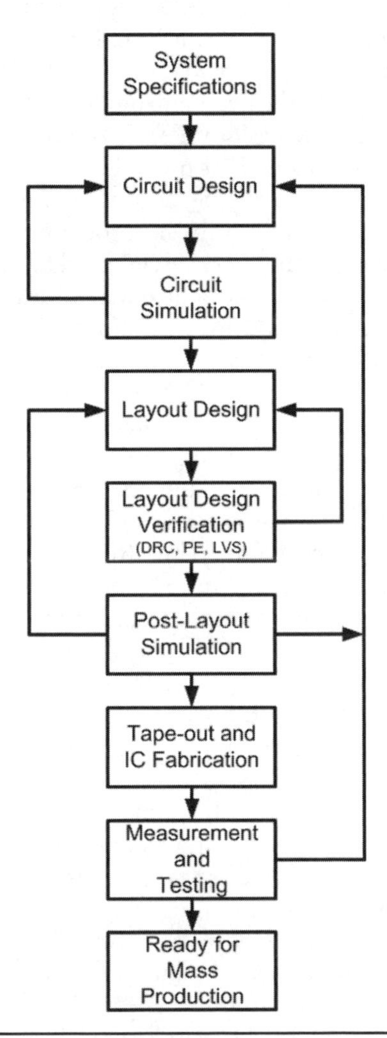

Figure 3-17 IC design flow chart

These specifications set forth the function to be performed by the IC and include design parameters such as operating frequency, speed performance, bandwidth, power consumption, gain, noise, and silicon area.

Once the specifications are set, the design process begins with a two-dimensional transistor-level circuit diagram/description, or circuit design in short, constructed using a CAD tool (e.g., Schematic Editor from the Cadence Virtuoso framework). Besides designing circuits in graphical form, circuit description may also be prepared in written format, commonly known as a netlist using a CAD tool such as HSPICE from Synopsys. The schematic diagram or netlist then undergoes circuit-level simulations (e.g., Cadence's Virtuoso Spectre or HSPICE circuit simulator) based on the model file or technology file from the foundry to make sure that the circuit operates correctly and its design parameters match the targeted specifications. The cardinal rules, which make up a *good* model file, are adequately simple to avoid excessive computer simulation time and capable to accurately predict experimental performance after fabrication. If the schematic design fails to meet any of the specifications, it has to be modified to optimize the performance of the circuit or sometimes redesigned so as to fulfill the design goals. Hence, the schematic design and circuit simulation steps keep on iterating until the circuit is fully functional and falls within the engineering specifications. An example of a schematic diagram drawn in the Cadence environment and a separate HSPICE netlist is depicted in Figure 3-18 and Figure 3-19, respectively.

To prepare the design for fabrication purposes, the next stage of the IC design flow is to convert the circuit schematic into a layout comprising the three-dimensional geometric artwork according to industry standards. This task is also carried out in a CAD environment, such as Cadence's Virtuoso Layout Editor. While drawing the layout, the designer must ensure that the layout adheres to a set of physical layout-design rules usually predetermined and imposed by the particular semiconductor foundry in which the layout will be fabricated. This is the first verification stage. Although the designer might perform a self-check during the graphical entry of the layout, there is still a possibility of overlooking any violations, especially in a complex circuit with an overwhelmingly large number of polygons that need to be checked against each other. Therefore, this routine verification step is often supported by simultaneously running an automatic design rule check (DRC) tool (e.g., Cadence's Assura and Mentor Graphics' Calibre DRC tool) to confirm that there is no violation of the design rules.[4] The set of design rules serves as a guideline that stipulates specific geometric and connectivity limitations such as the minimum dimension of width or spacing between metal layers. Strict conformance to these rules provides safe margins to minimize the impact of process variations that is a requisite requirement for high-yield and fault-free translation of physical layout

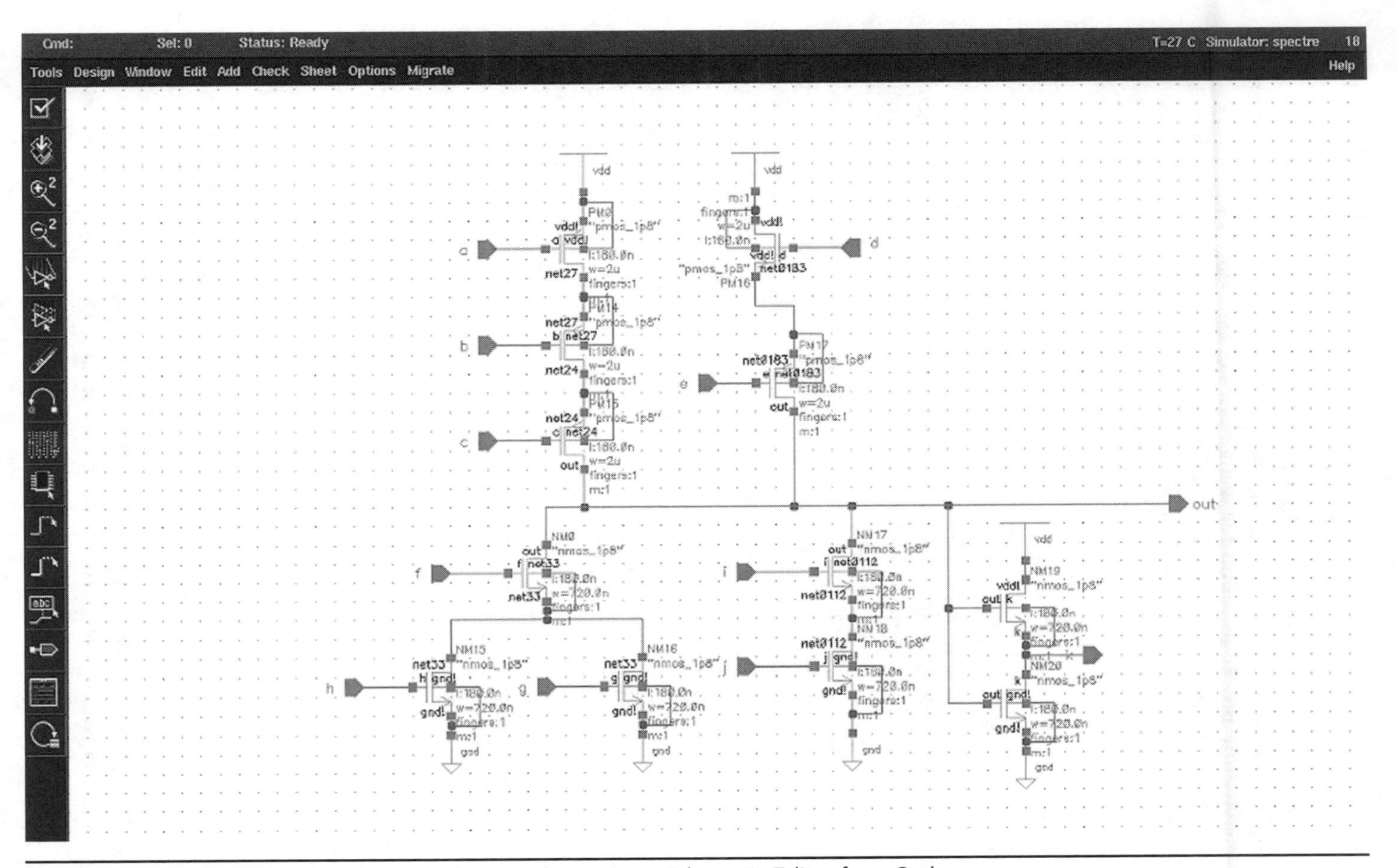

Figure 3-18 An example of a schematic diagram drawn using Schematic Editor from Cadence

```
* ADDER9.C.RAW
* NETLIST OUTPUT FOR HSPICES.
* GENERATED ON JUL 18 17:40:22 2008
* GLOBAL NET DEFINITIONS
.GLOBAL VDD!
* FILE NAME: ADDER_ADDER9_SCHEMATIC.S.
* SUBCIRCUIT FOR CELL: ADDER9.
* GENERATED FOR: HSPICES.
* GENERATED ON JUL 18 17:40:22 2008.
MNM20 K OUT 0 0  NMOS_1P8  L=180E-9 W=720E-9 AD=345.6E-15 AS=345.6E-15
+PD=2.4E-6 PS=2.4E-6 M=+1.00000000E+00
MNM19 VDD! OUT K K  NMOS_1P8  L=180E-9 W=720E-9 AD=345.6E-15 AS=345.6E-15
+PD=2.4E-6 PS=2.4E-6 M=+1.00000000E+00
MNM0 OUT F NET33 NET33  NMOS_1P8  L=180E-9 W=720E-9 AD=345.6E-15 AS=345.6E-15
+PD=2.4E-6 PS=2.4E-6 M=+1.00000000E+00
MNM15 NET33 H 0 0  NMOS_1P8  L=180E-9 W=720E-9 AD=345.6E-15 AS=345.6E-15
+PD=2.4E-6 PS=2.4E-6 M=+1.00000000E+00
MNM16 NET33 G 0 0  NMOS_1P8  L=180E-9 W=720E-9 AD=345.6E-15 AS=345.6E-15
+PD=2.4E-6 PS=2.4E-6 M=+1.00000000E+00
MNM17 OUT I NET0112 NET0112  NMOS_1P8  L=180E-9 W=720E-9 AD=345.6E-15
+AS=345.6E-15 PD=2.4E-6 PS=2.4E-6 M=+1.00000000E+00
MNM18 NET0112 J 0 0  NMOS_1P8  L=180E-9 W=720E-9 AD=345.6E-15 AS=345.6E-15
+PD=2.4E-6 PS=2.4E-6 M=+1.00000000E+00
MPM16 VDD! D NET0183 VDD!  PMOS_1P8  L=180E-9 W=2E-6 AD=960E-15 AS=960E-15
+PD=4.96E-6 PS=4.96E-6 M=+1.00000000E+00
MPM14 NET24 B NET27 NET27  PMOS_1P8  L=180E-9 W=2E-6 AD=960E-15 AS=960E-15
+PD=4.96E-6 PS=4.96E-6 M=+1.00000000E+00
MPM15 OUT C NET24 NET24  PMOS_1P8  L=180E-9 W=2E-6 AD=960E-15 AS=960E-15
+PD=4.96E-6 PS=4.96E-6 M=+1.00000000E+00
MPM17 OUT E NET0183 NET0183  PMOS_1P8  L=180E-9 W=2E-6 AD=960E-15 AS=960E-15
+PD=4.96E-6 PS=4.96E-6 M=+1.00000000E+00
MPM0 NET27 A VDD! VDD!  PMOS_1P8  L=180E-9 W=2E-6 AD=960E-15 AS=960E-15
+PD=4.96E-6 PS=4.96E-6 M=+1.00000000E+00

* INCLUDE FILES

.lib "/net/wildar/software2/local/library/CSM/CSM018PDK/csm18rf/../models/sm093001-1j.scs" section=typical
.lib "/net/wildar/software2/local/library/CSM/CSM018PDK/csm18rf/../models/sm093001-1j.scs" section=resistor
.lib "/net/wildar/software2/local/library/CSM/CSM018PDK/csm18rf/../models/sm093001-1j.scs" section=bjt
.lib "/net/wildar/software2/local/library/CSM/CSM018PDK/csm18rf/../models/sm093001-1j.scs" section=diode
.lib "/net/wildar/software2/local/library/CSM/CSM018PDK/csm18rf/../models/sm093001-1j.scs" section=capacitor
.lib "/net/wildar/software2/local/library/CSM/CSM018PDK/csm18rf/../models/sm093001-1j-mvt.scs" section=typical
.lib "/net/wildar/software2/local/library/CSM/CSM018PDK/csm18rf/../models/sm093001-1j-5p0.scs" section=typical
.lib "/net/wildar/software2/local/library/CSM/CSM018PDK/csm18rf/../models/sm093001-1j-5p0.scs" section=diode
.lib "/net/wildar/software2/local/library/CSM/CSM018PDK/csm18rf/../models/sm093009-1b.scs" section=typical
.lib "/net/wildar/software2/local/library/CSM/CSM018PDK/csm18rf/../models/sm093009-1b.scs" section=diode
.lib "/net/wildar/software2/local/library/CSM/CSM018PDK/csm18rf/../models/twell.scs"
.lib "/net/wildar/software2/local/library/CSM/CSM018PDK/csm18rf/../models/ind.scs"
.lib "/net/wildar/software2/local/library/CSM/CSM018PDK/csm18rf/../models/ind_TM.scs"
.lib "/net/wildar/software2/local/library/CSM/CSM018PDK/csm18rf/../models/mim_analog_rf.scs"
.lib "/net/wildar/software2/local/library/CSM/CSM018PDK/csm18rf/../models/pn_var.scs"
.lib "/net/wildar/software2/local/library/CSM/CSM018PDK/csm18rf/../models/mos_var.scs"
.lib "/net/wildar/software2/local/library/CSM/CSM018PDK/csm18rf/../models/rfmos.scs"

* END OF NETLIST
.TRAN  1.00000E-11 4.00000E-09 START= 1.00000E-09
.TEMP    25.0000
.OP
.save
.OPTION  INGOLD=2 ARTIST=2 PSF=2
+        PROBE=0
+        DELMAX = 5.00000E-11
.END
```

Figure 3-19 An example of an HSPICE netlist

into actual geometry during the IC manufacturing process. An example of a layout-design for an arbitrary circuit design drawn using Cadence Virtuoso is shown in Figure 3-20.

Now that we are assured that our layout is DRC-clean and ready for fault-less fabrication, you might have these questions: How do we tell whether the drawn layout is really equivalent to the original circuit topology intended for fabrication? How do we know if there is any missing or possibly unintended

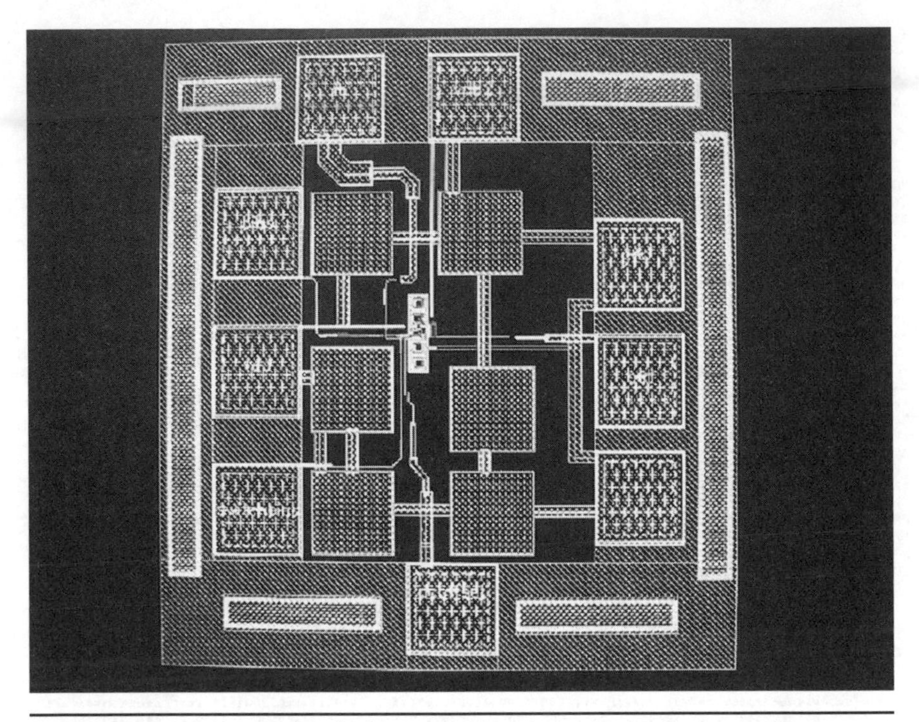

Figure 3-20 An example of a layout-design drawn in the Cadence Virtuoso environment

connection between devices? Are the layout and schematic topologically matched? Clearly, this is why a second verification stage—layout versus schematic (LVS) check—is needed. This is a correlation process using the built-in verification feature of a CAD tool (e.g., Assura LVS checker from Cadence) to determine whether a particular layout correctly corresponds to the original circuit diagram of the design. However, take note that prior to the LVS check, we must first perform a parasitic extraction (PE) step. The PE tool reads the physical data of the layout and identifies transistors, interconnects, and any other circuit elements based on the various polygons and mask layers. It also calculates the parasitic capacitances and resistances that are intrinsically attached to these structures, and eventually creates a detailed circuit description (also known as netlist) that gives an accurate estimation and description of the device dimensions and parasitics in the layout. This *extracted netlist* is used to compare against the circuit schematic netlist in LVS verification and subsequently used in layout simulation.

In essence, the LVS step provides an additional level of confidence for a reliable correlation between the layout and the circuit we want to fabricate. Any errors that may be prompted during the LVS run should be rectified accordingly by modifying the layout and/or circuit schematic and rerunning the LVS check. This correlation process continues until the layout and the schematic topologies are perfectly matched. Please bear in mind that the extraction step has to be repeated whenever the layout is changed. Once the verification step of the generated layout is completed (DRC and LVS clean), we can proceed to post-layout simulation.

Post-layout simulation is performed by designers to evaluate the electrical performance of a full-custom IC design based on the extracted netlist. As compared to circuit-level simulation, post-layout simulation provides a more precise assessment of the design metrics of the designed circuit as parasitic effects are taken into consideration. If the post-layout simulation results do not comply with the design specifications, designers will have to modify the layout or transistor dimensions, or at times there may be a need to devise an improved circuit design, depending on the degree of deviation from the desired specifications. This is again an iterative process that requires concentrated and meticulous effort from the designers.

If the design meets all the specifications, we can move on to send the design file, typically in the form of GDS-II* stream format, for manufacture. This process is widely referred to as design *tape-out* in technical terminology.[15] Tape-out is an achievement which usually calls for celebration for the team of IC designers involved in the design process (you can start to plan for your holiday getaway now!). The IC fabrication procedure normally takes three to six months, depending on the foundry. Note, good post-layout simulation results do not necessarily warrant a successful product. In reality, the actual performance of the IC can only be ascertained by carrying out measurement and testing of the IC hardware returning from the fabrication facility. This measurement and testing step may be carried out at wafer level and/or package level. Depending on the requirement of the IC, such as the input/output pin assignment or complexity, the bare die may need to be packaged or encapsulated before chip evaluation. If the measurement result is not on-target even at the wafer level, the designers need to perform troubleshooting or debugging of the IC to identify potential technical flaws. All design specifications must be met before proceeding to the packaging step. As mentioned briefly in Chapter 1, there are a number of significant advantages of encapsulating the IC. Indeed, the IC packaging technology improves the reliability of the IC from the surrounding environment by providing electromagnetic shielding and protecting it from corrosion. Furthermore, since the minute size and delicate nature of the silicon die makes it a challenge to mount the IC on the PCB and other components that ultimately form the electronic devices for consumers' usage such as microcomputers, electronic watches, and pocket calculators, the encapsulation provides the IC with a structure to connect to the external world. In addition,

---

*Short for graphic data system, GDS-II is an industry standard format of literal encoding that describes the way a layout is designed using words and numeric attributes

besides optimizing the power consumption and speed performance of an IC, the encapsulation has metal pins that provide a means of electrical connections to bridge the signals in and out of the packaged die.

In the measurement and testing step, the fabricated IC prototype or the packaged IC is subject to rigorous electrical and functional measurement and testing on top of specification checks. A photomicrograph of a wafer die under test is shown in Figure 3-21. This is the critical moment every designer eagerly awaits and anticipates: is it a success or a failure? Based on the measurement and testing results, the IC chip is either released for mass production (Bravo!! Go reward yourself! You can go on holiday now!) it or reenters the appropriate step of the design cycle (Sorry, you might have to delay your holiday plan).

From the above explanations, we can see that IC design is a challenging task—both intricate and tedious. Depending on the scale and complexity of the design, a substantial amount of time and constant intellectual effort from a dedicated team of IC designers goes into designing, testing, checking, and ultimately perfecting (in an ideal case) the IC. This justifies the rationale and importance of protecting the layout-design of an IC by using the intellectual property framework.

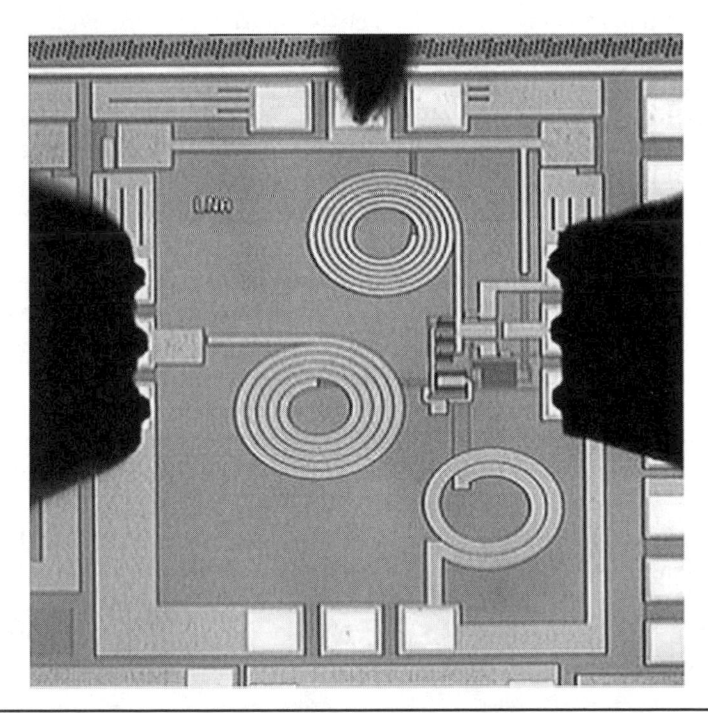

Figure 3-21 A photomicrograph of a wafer die under test

Summary

We have presented a bird's-eye view of the technical aspects of integrated circuits (ICs) in this chapter. The actual theoretical background of ICs is, of course, much more profound and complicated, but our goal here is to portray the *big picture* of the technical framework of the IC field using a nontechnical approach. This is to prepare our nontechnical audience with relevant IC knowledge so as to understand the concept of intellectual property (IP) for ICs in the remaining chapters.

We began with a definition of the IC and the key motivation behind the invention of ICs. This was followed by an overview of the evolution of the IC fabrication technology since the emergence of the IC industry that includes process technologies such as bipolar, CMOS, and BiCMOS. A separate section devoted to describing IC applications imparted the various classifications of analog, digital, and mixed-signal ICs. The common types of transistors available in the IC industry today, MOSFET and BJT, were also explained. The definition of semiconductor and the essential concept of doping and *p-n* junction were also covered. In addition, we discovered that the feature sizes of the transistors have been decreasing drastically as evidenced by the continual advancement in successive generations of technology nodes and the increased level of device integration on a single chip. In view of this, design automation has been envisioned as a viable solution and has indeed been proven paramount to address skyrocketing circuit complexity while circumventing the narrowing time-to-market window. The flow of IC design, which elaborates the steps associated to designing an IC from the beginning to its production stage, concludes the discussion in this chapter.

This book has free material available for download from the Web Added Value™ resource center at *www.jrosspub.com*

4

Types of Intellectual Property for Integrated Circuit Protection

Various forms of IP are available in the legal protection regime that are intended for the protection of different classifications of human intellectual creations. In fact, there are substantial variations across international boundaries in terms of the procedure for obtaining a particular form of IP, the requirements placed on the patentee, the duration of protection, and the extent of rights being conferred according to national laws. How a certain form of IP is named may also be different from nation to nation. The typical forms of IP which are widely recognized are: patent, copyright, layout-design for integrated circuits, trademark, industrial design, trade secret, plant variety, and geographical indication. These IP vehicles significantly differ in the area they cover, the rights they confer, the procedure for obtaining them, and how they are maintained.

Nearly all of the IP forms are relevant or applicable in one way or another to the specific area of integrated circuits, except plant variety and geographical indication. To equip our readers on their route to take legal precautions to protect their intellectual innovation, it is essential for us to acquaint you with the various forms of IP for ICs, thereby enabling you to identify and determine the type of IP that is appropriate for your specific needs. In this chapter, we present a detailed explanation of each type of IP relevant to the IC industry and discuss when and how they are used.

Patents

A patent (or utility patent as it is known in some countries, notably the United States) is an exclusive monopoly right issued by a government to the owner (patentee) of an invention for a limited period of time. It gives the owner the sole right to forbid others from using, copying, making, offering for sale, selling, or importing the invention without his or her prior consent in the country in which he or she obtained patent protection. It follows that patent is territorial in extent. In other words, protection does not normally extend beyond the borders of the country in which the patent is granted.

Nature of Patent Rights

A somewhat confusing point—but nonetheless essential—worth emphasizing here is that the patent right is of a negative exclusionary nature. The conferment of a patent *does not* grant its owner the right to make, use, offer for sale, sell, or import the invention if any of those acts is in violation of the laws within a country. Rather, the patentee is strictly conferred purely the right to *exclude others* from doing so. The *patentee's own right to utilize the invention* is therefore independent of the patent granted. Does this sound ironic? You will understand the rationale behind it after following through our explanation. Let us elaborate on the concept with some examples.

Numerous new inventions are incremental or improved derivations developed based on their predecessors that may be protected by a patent owned by the earlier inventor. Therefore, even though a patent may have been obtained for a new invention, its manufacture and sale can still infringe upon a prior patent owned by another inventor. For instance, inventor *Jonathan* (names used are anonymous) holds a patent of an improved mobile phone design with an integrated camera feature that is based on an existing design of the mobile phone with basic functionalities covered by a patent owned by *David*. However, the patent that *Jonathan* is holding does not authorize him to freely exploit his invention to its fullest potential. Whenever *Jonathan* manufactures or sells his new mobile phone with an integrated camera, he will avoidably step into the rights to exclude held by *David*'s original patent. Hence, *Jonathan* can only legally make or sell (or engage in all other commercial activities) his improved version of the mobile phone by acquiring prior permission from *David*, who is the patent holder of the original mobile phone—provided the original patent is still in force. Conversely, the patent held by *Jonathan* allows him to exclude *David* from exploiting the mobile phone with a built-in camera feature. Therefore, a patent only conveys an absolute prohibitive right that serves to prevent third parties from exploiting the invention. Meanwhile, the patentee's own right to leverage his or her invention is restrictively subjected to the rights of others and any general laws that might be applicable to ensure that the product is legal and safe for use.

Patentable Subject Matter

For an idea or invention to be rendered patentable, it must first fall within the categories of patentable subject matter. There are some forms of subject matter that are deemed exclusionary from the realm of patent eligibility. As always, depending on respective national legislation, the rules governing the patentability of subject matter differ quite substantially from one country to another. In general, new products or processes that give technical solutions to problems are considered patentable subject matter. This includes novel methods of solving problems, new machines (devices or apparatuses that have moving parts), compositions of matter (combination of chemical compounds or other materials that achieve a certain result, such as drugs), or incremental technical improvements of certain devices. A subject matter is not patentable if it is a discovery, a manifestation of the laws of nature, an abstract idea, or an aesthetic creation. For example, Isaac Newton could not patent his much celebrated discovery of the law of gravity, and neither could Albert Einstein patent his theory of relativity as these are all theories about the laws of nature, which no man can call exclusively his own. However, inventions applying these theories can be held patentable. Moreover, aesthetic or artistic materials such as literary works, musicals, dramas, or mere compilations of data without functionality are not patentable.

Some other subject matter that are commonly barred from the scope of patentability in most countries include methods of medical treatment of the human or animal body by means of surgery or therapy or diagnostics and inventions that are generally expected to instigate elements of public disorder such as morally destructive or antisocial behavior. The patentability of subject matter such as business methods, mathematical algorithms, software, and life forms is particularly controversial from a global perspective, which has incited heated battles in recent years.[1 3] Generally, the U.S. definition of patentable subject matter is perhaps the most liberal in the world whereas other countries are more conservative in regard to the issue.

Patentability

For an invention or subject matter to be patentable, it must typically comply with several requisite requirements (also known as patentability tests). In fact, patent systems globally use different terms for these concepts and impose varying standards for determining patentability. The most widely used and prominent are presented in the following paragraphs.

Novelty or Newness: The invention must be novel and new. It must not be an exact resemblance of the prior art. It should be accentuated here that prior art is not limited to patents. It embraces all information, including prior products or processes that can be found in written fixed mediums or publications, or already in use or on sale in the marketplace. The said invention never should have been disclosed to the public in any way or anywhere before the date on which a patent application is made. The novelty feature can be negated if it has already been exposed to the public in any way and includes research articles or magazine publications, press advertisement, public demonstration or use, or simply by word of mouth. Pursuant to this, the onus is on the inventors to exercise vigilance to keep the invention secret until the application for a patent is successfully filed. It is also important for inventors to practice extra care and prudence through implied confidentiality should there be a need of disclosing the invention to engage external support, such as marketing personnel or potential investors, before the patent filing date.

The absolute sense of novelty is not a universal rule. In fact, there are important exceptions whereby there exists a grace period which allows for invention disclosure without affecting its patentability in countries such as the United States (one year), Japan (six months), Canada (one year), and Australia (one year), as defined in their respective national patent laws. The availability of the grace period remedies the dire consequences of inadvertent, impulsive, or inconsiderate divulgement of the invention to the community at large, without which the inventor would find himself in a state of complete and permanent loss of all patent rights in the invention.[4] A grace period provides a specific timeline within which a subsequent patent application may be filed despite prior disclosures of the invention to the public anywhere around the world. Therefore, the novelty status is preserved and a patent may still be granted on the condition that all other formalities and legal requirements are complied with for patentability. However, it is crucial to recognize the first public disclosure, as it triggers the grace period clock to start ticking. After the grace period, an inventor, failing to file a patent application for the new invention, will have to concede all patent rights for the invention.

Inventive Step or Non-Obviousness: Besides being novel, the claimed invention must also involve inventive ingenuity. It must be nonobvious in light of the prior art and go beyond the routine expertise, accumulated knowledge, and past experience of an ordinarily skilled person who is familiar with the particular technological field of the invention. In other words, the invention must exhibit a considerable enhancement when compared to the state-of-the-art before it would progress to the stage qualifying for patent protection. Pertaining to this requirement, even if an invention is novel yet obvious and intuitive to a person skilled in that field, the invention does not satisfy the inventive step patentability test. Some examples of changes to known products that do not normally qualify for patentability are simple substitutions of one known material with another or size alteration, among others.

Ambiguities and hurdles often arise when patent applicants are confronted with whether their present inventions are sufficiently inventive or nonobvious to deserve patent monopoly. This is because there is no lucid definition of what an obvious invention is and frequently, therefore, results in an unavoidable judgment call. As a result, due to its inherent obscureness, the inventive step or nonobviousness requirement is widely considered the most intricate obstacle for a patent applicant to overcome and often a grey area provoking controversial debates in the process of determining the validity of a patent.

Industrial Applicability or Utility: The invention must be useful and hence susceptible or capable of practical industrial application. In other words, it should be capable of being made or used, for achieving a concrete and substantial end result in any industry, or simply put, economically viable.[5] For example, an inventor who proposes an invention of a new groundbreaking machine, albeit one that does not operate to perform its intended function or is incapable of yielding identifiable benefit, will fail in his or her patent application endeavors. In general, the term *industry* is commonly interpreted in the broadest possible sense, which includes agriculture.[6] In most cases, the industrial applicability or utility requirement is easily accounted for in IC-related technologies.

Term of Patent

The term of a patent is the maximum period of time during which it remains enforceable or protected against infringements. In most countries, the term for patent protection is 20 years from the first patent filing date or patent grant date, and is further subjected to regular payments of maintenance fees to keep the patent in force. The purpose of introducing maintenance fees is to encourage patent owners to periodically assess the economic worth of their patents. Failure to make payment on time may result in having the patent forfeited before its term. Upon actual expiration of the patent term, it might still be extended under special circumstances, such as delayed issuance of a patent. Once the patent ultimately expires, anyone from the general public would have the full entitlement of using that particular invention without having the concern about infringing on the rights of the patentee.

Copyright

Copyright, as its name implies, is basically the right to copy an original work. It applies to a wide range of literary, musical, dramatic, artistic, and other creative works. It is a form of IP granting its owner, in many cases the creator or author of the work, the bundled rights to use the work in question for his or her own intended purposes. Included in that bundle are rights of reproduction, derivation (translation, abridgment, or adaptation), public distribution, and performance and display of the whole or a substantial part of the copyrighted work. In the meantime, the copyright owner can authorize another party of any of these activities, or, on the contrary, prohibit others from engaging in any such acts, entirely at his or her discretion. If copying is authorized, the copyright owner has the right to be further credited for the work or associated to the work by name. Nevertheless, unlike patent, copyright provides no protection to its owner if a similar work is developed independently by a third party without copying the original work. Furthermore, mere possession of the copyrighted work does not give a person the rights attached to the copyright.

Like patent, copyright is enacted by the majority of governments from all globally. In most countries, an author need not apply or register for copyright protection for his or her work because copyright entitlements come in force automatically once the intellectual work becomes tangible. As such, the use of copyright notice, which includes all rights reserved and copyright symbols (© for visually perceptible works and ® for sound recordings) is not mandatory to assert copyright, but it is nevertheless a good practice and serves as a reminder to the public to respect the rights pertaining to copyright law. Similarly, registration is optional, but in countries where the registration procedure is available, it is beneficial to attain maximum protection while enforcing rights, including the establishment of the evidence of a valid copyright claim and enabling the copyright owner to file an infringement lawsuit to seek monetary damages and recover reasonable attorney fees.

Subject Matter of Copyright

Copyright covers the aesthetic aspects of creative works. It protects nonutilitarian or nonfunctional articles, which include literary, musical, dramatic, artistic, and other intellectual creations. In other words, it protects objects which exist simply because of their exterior appearance and not for their usefulness.[7] Some examples of each category of copyrightable subject matters follow:

Literary: computer programs, databases, speeches, poetry, stories, manuscripts, books, song lyrics, and media, e.g., newspapers, magazines, and the internet.

Dramatic: choreography (dance movements), pantomimes (acting using bodily movements, gestures, and facial expressions), plays, films, videos, and scripts.

Musical: compositions that are comprised of both music and lyrics, or music only.

Artistic: graphic expressions, drawings, paintings, photographs, charts, engravings, etchings, works of craftsmanship (sculptures), and architectural works (buildings and models of buildings).

It is also important to note that as specifically stated in copyright law, no protection is given to ideas, concepts, information, principles, discoveries, methods of operation, processes, or systems, notwithstanding the form in which it is described, explained, illustrated, or embodied in such work. In addition, short phrases such as names, titles, and slogans are not covered under the copyright regime.

Copyright-ability

Now we understand the scope of intellectual work in which copyright protection subsists. Additionally, there are two pertinent requirements for a subject matter to satisfy the criteria for the copyright regime. These requirements are:

Embodiment in a tangible medium of expression: Copyright protects only the expressible form of an idea or information, or the way the idea or information is embodied. The protection does not extend to the underlying idea or information itself. Hence, an idea has to be reduced to a physical representation, or in other words, *fixed* or recorded on some tangible embodiment, which includes recordings, drawings, canvas, writings, letters, audiocassettes, computer software, compact discs, etc. For example, the facts and news reported in a newspaper article are not protected. Only the expression (in this case in text form) in which the facts or news are reported is protected under copyright law. Another example pertains to Disney's widely loved cartoon characters such as Mickey Mouse—its copyright merely prevents others from copying or adapting the form of expression of Disney's particular anthropomorphic mouse but does not impose restrictions on other creations of mice in general—on the condition that they are substantially different from Disney's expression. Therefore, copyright does not prevent others from using the idea or concept that is embodied in the copyright owner's work. This implies that anyone is free to develop their own expression of precisely the same idea with impunity.

Originality: It is not necessary for a work to be novel to attract copyright protection (unlike patent), nor must it be aesthetically appealing to be copyrightable. Essays written by high school sophomores are as equally copyrightable as J. K. Rowling's hugely popular Harry Potter novels. However, one cardinal rule to remember is, for the work to deserve copyright protection it must be original, which means that the expression was not copied from another prior source but was developed independently by the author.

Term of Copyright

Contrary to physical properties, such as a toy, a television set, or a house, in which ownership is perpetual until they are sold, consumed, or given away, the legal term of intellectual properties terminates at a certain point in time. There is no exception to copyright law. The term of copyright varies depending on the categories of copyrighted work and different jurisdictional laws. Generally speaking, for works created by individuals, the term of copyright protection is the life of the author plus an additional 70 years after the demise of the author. If the work was created by a corporation, the term is 95 years commencing from the first publication of the work. Upon expiration of the effective period of the copyright, the material is returned to the public domain and anyone shares an equal right to use or exploit it without the need for requesting copyright permission or clearances. For example, William Shakespeare's works are part of the public domain following his death in 1616—copyright has been lifted and everyone is free to adapt, perform, or produce them.

Layout-Designs of Integrated Circuits Act

The legislation for layout-designs of ICs (in short, IC layout) protection is a comparatively new breed of IP law, which is in many aspects a hybrid across the patent and copyright domains.[8] We detail this form of IP, because it is an enactment customized specifically for protection of IC layout design. Registration for protection may be required depending on the legislation of a country. The protection lasts between 8 and 15 years counting from the filing date of application for registration, or from the first commercial exploitation of the layout-design, based on the legal systems of the respective countries. The shorter duration of protection for layout-designs when compared to those of the patent and copyright systems reflects the fast pace of innovation that is taking place in the IC design industry.

Historical Facts

The country that first initiated a statute for IC layout-designs protection was the United States, the place where the IC was born. It was known as the Semiconductor Chip Protection Act (SCPA), enacted in 1984. It is akin to the Copyright Act in many significant ways. The similarities will become evident as we progress through this section. The enactment is commonly known as the sui generis regime for ICs, which applies specifically to the semiconductor industry. The phrase *sui generis* is a Latin expression. By definition, it means *only one of its kind*, or *unique in its own characteristics*.[9]

In the mid-1980s, the SCPA became the mold on which other nations lay their foundations for the legal protection of IC layout-designs. In 1989, the World Intellectual Property Organization (WIPO) introduced the Treaty on Intellectual Property in Respect of Integrated Circuits, aka the Washington or IPIC Treaty,[10] in an attempt to harmonize the laws internationally. However, this treaty did not come into force due mainly to disagreements from the United States and Japan, the world's chip powers during that era. The Agreement on Trade-Related Aspects of Intellectual Property Rights (TRIPS) was negotiated and signed in 1994.[11] This international agreement contains, among other forms of IP, the Layout-Designs (topographies) of Integrated Circuits Act, which incorporates remedies, modifications, and new additions to the earlier treaty's provisions. In essence, the provisions related to the protection of layout-designs of ICs under TRIPS are similar to its Washington treaty's predecessor except for some important adaptations. All signatories to the World Trade Organization (WTO), which include virtually all countries in the world, are obliged to protect IC layouts in accordance with the provisions of the TRIPS agreement.

The legislation for IC layout protection is named or phrased differently in many member countries but fulfills obligations to the TRIPS agreement. To cite some examples, they are named the Integrated Circuit Topography Act in Canada, Layout-Designs of Integrated Circuits Act in Singapore, Integrated Circuit Layout Protect Act in Taiwan, and Circuit Layouts Act in Australia—all fundamentally similar in concept but with some varying scope and degree of protection. In our discussion, we use a more generalized approach for the reader to understand the gist of this particular form of IP. Hereafter, we shall use the term *the Act* to represent this sui generis protection regime.

Need for a Sui Generis Protection for IC Layout

The effort of designing IC layouts is usually brought to fruition with enormous expertise, creativity, and capital investment.[12] Paradoxically, when the products enter the marketplace, each piece of the product carries with it its own *treachery* blueprint of how to make a clone.[13] Hence, these works are exceptionally susceptible to easy, rapid, and competitive misappropriation by technological means,[8] which often require only a tiny proportion of the developers' cost. In view of the aforementioned, protection of IC layouts is of ultimate importance.

There is a common perception that the scope of protection for IC layout-design is generally a poor fit into either of the two dominant IP paradigms, namely patent and copyright. This is due to the fact that an IC design is likely to fail the patent law's criteria of nonobviousness, and its inherent utilitarian or functional nature would essentially exempt it from copyright protection.[14,15] It was due to this nonutilitarian precept of copyright law that has thwarted Intel's application for chip protection via copyright in its battle against so-called *chip piracy* since the late 1970s.[16] Furthermore, copyright law may be too general, hence inadequate to embrace the scientific know-how of IC layouts.[17]

The establishment of a sui generis form of IP law following a legislative reform was perceived as a radical move by the U.S. government. The normal practice was to virtually amend existing patent or copyright laws by integrating the subject matters into either one of them rather than substantiating a wholly unique form of IP right.[18] However, it was a difficult challenge to amend the existing laws to accommodate layout-designs of ICs as that would mean the need to forsake the underlying tenets of patent and copyright regimes. On the other hand, IC designers cannot afford to let competitors exploit their designs which are often the result of an extensive amount of effort, time, and money, depending on the level of complexity. In a nutshell, the specific and immediate needs of the rapidly growing IC design industry and the lack of an appropriate legal regime have motivated the bold move of tailoring a separate law scheme for IC design.

Subject Matter of IC Layout Protection

The Act essentially protects a layout-design of an IC against unauthorized exploitation. Also known as IC topography, IC layout is defined as the "three-dimensional disposition, however expressed, of the elements, at least one of which is an active element, and of some or all of the interconnections of an integrated circuit, or such a three-dimensional disposition prepared for an integrated circuit intended for manufacture."[10] Reduced to its essence, IC layout refers to the predetermined design of three-dimensional, layered pattern of elements and interconnections of an IC. Even though the term *mask work* has been used to represent the protected subject matter under the U.S. SCPA, it is apparently not a good idea because its technology-specific nature would render the SCPA obsolete. ICs may be fabricated by means of manufacturing techniques that evolve beyond the specific use of masks in the chip fabrication process.[13,16]

On the other hand, in a close resemblance to copyright law, protections under the Act also do not extend to ideas, concepts, information, principles, discoveries, methods of operation, processes, or systems, notwithstanding the form in which it is described, explained, illustrated, or embodied in a layout design.

Eligibility for IC Layout Protection

To be eligible for protection under the Act, a layout-design must abide by the following statutory tests:

Originality: The layout-design must be original, which means that it is developed by the intellectual endeavor or research effort of its creator and is not simply a reproduction of the entire or significant portion of another preexisting layout-design. Examples of evidence to show that intellectual effort has been elicited are drawings of the layout-design, documentation, etc. Moreover, the work must not be deemed commonplace, familiar, or staple among fellow layout-designers and IC manufacturers at the time of its creation.

As for a layout-design that comprises a combination of elements and interconnections that are commonplace, the combination, considered as a whole, must be original in the same context as stipulated earlier for it to warrant protection under the Act. Additionally, if there is enough evidence to prove that a layout-design has been created independently, it is considered fulfilling the originality requirement, even if it is identical to another piece of protected design.

Reduction to Expressible Form: As discussed earlier, intangible or abstract ideas are not subject matter for protection under the Act. Therefore, any idea of an IC layout shall be viewed as not existing at all until it is converted into a documentary or physical form. The material forms within which the conceptual creativity of an IC designer can assume shape are photolithographic mask, circuit schematic diagram, layout data (e.g., GDS-II* stream format), layout-design drawing, photograph, or embodiment in the actual IC.[14] Hence, it is generally not a precondition for the layout-design to be incorporated in an IC product for it to be protected under the legislation of most countries.

Exclusive Rights Granted

Under the Act, the qualified owner of a layout-design is granted the exclusive statutory right to copy or reproduce the protected layout-design in its entirety or any part of it, whether by embodiment into an IC (be it in its final or intermediate form) or otherwise. The rights are similar to that of copyright. The rights holder is given the authority to commercially exploit the protected layout-design or an IC product in which all or any part of the layout-design is embodied. Commercial exploitation includes import, sale, lease, offering or displaying for sale or lease, or any other means of distribution for trading purposes. In addition, the rightful owner of the

---

*Short for graphic data system, GDS-II is an industry standard format of literal encoding that describes the way a layout is designed using words and numeric attributes

layout-design has the right to authorize or prevent a third party from engaging in any of the above-mentioned acts. However, unlike copyright, the Act does not grant rights to the owner of a protected layout-design of a derivative work.[16] An important exception to note is that the Act allows copying or reproduction of the protected layout-design or any part of it if it is meant solely for private use (not for commercial purpose), research or teaching, or for analyzing or evaluating the concepts embodied in the topography. More details on this topic are provided in Chapter 6.

Industrial Design

Industrial design refers to the ornamental manifestation or aesthetic appearance of a utilitarian article. It may be embodied in, or applied to, the entire or only a part of the article.[19] On top of fostering valuable brand identity via product recognition, industrial design can be used strategically to increase the commercial or brand value of a product that also greatly enhances its marketability.[20,21] Industrial design law is employed mainly to protect such novel designs with industrial applicability. More precisely, it is the right given by a government to the creator of a new, original, and ornamental design for an object or article of manufacture to have control over, or to profit from the use of his or her invention. The owner of an industrial design is conferred the right to exclude a third party from using, reproducing, or selling the design or lookalike imitations without his or her consent. This is important to instill creativity while promoting the invention of more products that are aesthetically pleasing.

Registration is necessary in most jurisdictions to enjoy protection under industrial design law. In some countries, such as the United States, the legislation for industrial design is classified under the main IP regime of patent and is known as the design patent. In contrast to utility patents that are protected for a period of up to 20 years from the filing date, U.S. design patents are valid for a 14-year nonrenewable term from the date of issuance of the design patent. In most other legislation, the term of protection is normally five years from the date of the grant of the design patent with the possibility of further renewals of up to 15 years in the majority of cases.[21]

Subject Matter for Industrial Design Protection

Industrial design is basically what makes a utilitarian object appear appealing to the eyes. Thus, since industrial design is essentially of an aesthetic nature, its protection does not in any case extend to the structure or utilitarian aspects of these articles. Put simply, industrial design rights only protect the ornamental uniqueness of a utilitarian article, but do not protect the utilitarian feature of the article. This is in contrast to the copyright that covers nonutilitarian articles and the patent that protects the utility or technical aspects of an article.

The subject matter within the scope of industrial design protection include the features of shape, configuration, surface ornamentation, or pattern applied to the article by an industrial or manufacturing process. For instance, the design of garments, automobiles, printers, perfume bottles, jewelry, wrist watches, computer monitors, keyboards, and CPU case—all may be protected by industrial design rights. There are inventions which embrace both ornamental and functional characteristics. For such inventions, industrial design protection alone is not sufficient—rather, patent (or utility patent in some countries) comes into the picture as well. Therefore, for the same invention, two types of protection subsist: industrial design protection for the ornamental aspect of the invention and utility patent for its functional aspect.

In another case, if the design itself is both ornamental and utilitarian in nature, then the industrial design right does not prevail in this respect.[22,23] Rather, the design can be conferred utility patent protection. Examples of designs residing in this category are, among others, enhanced shape designs of a computer mouse or computer keyboard based on ergonomic principles that give more comfort to the user, or the unique anterior design of a car that results in reduced wind resistance. Therefore, a design for an article of manufacture is considered an improper subject matter if it is dictated primarily by the function of the article (with sole utilitarian motives) and lacks ornamental distinctiveness.[22] Frequently, the ornamental and utility features of an article are not easily distinguishable. In addition, designs that incur public disorder or immorality are also excluded from protection under industrial design.

Registrability of an Industrial Design

Like most other types of IP, the cardinal rule for an industrial design to deserve registrability is originality or newness. This means that no identical or close resemblance of the design can exist before the first filing date of application for protection. It must not have been registered in the particular nation where protection is being sought or anywhere else. Therefore, it is the responsibility of the creator of a design to secure his or her design in confidentiality until a design application is filed.

Other than the originality requirement, the design must also be applied onto an article by an industrial process. To meet this requirement, the law of some countries, such as Singapore, provides that at least 50 replicas of the article or product bearing the industrial design must have been, or are intended to be, manufactured for sale, hire, or other means of commercial engagement.[24]

Trademark

A trademark is a distinctive sign, or a combination of signs, which is adopted and used by an individual, merchant, or manufacturer in the course of business or trade to identify and distinguish goods or services from those offered by other traders. It also serves as an indicator of source or commercial origin of its products to prospective consumers. A trademark used in relation to services rather than products is a subset of a trademark called a service mark.[25] Service marks are usually used by business entities in the service industry such as hotels, tourist agencies, restaurants, and airlines to identify or distinguish the services they provide.

Subject Matter of Trademark

The subject matter that qualifies for registration under trademark can be any letter, word, name, signature, numeral, logo, picture, device, brand, heading, label, ticket, aspect of packaging, or a combination of any of these. Over the years, trademark has seen numerous changes in the form of subject matter that can be protected and in some specific situations the protection has extended beyond the traditional array of subject matter. An increasing number of countries allow for the registration of less-conventional types of trademark such as appearance (color), three-dimensional signs (shapes or product packaging), audible signs (sounds), and olfactory signs (smells).[25,26] The registration of such subject matter is often a challenge, but may, nonetheless, be successful so long as they are unique and are capable of distinguishing the products or services from other competitors.

Most of us have watched movies at least once that are preceded with a roaring lion. The lion is a well-known logo associated with Metro-Goldwyn-Mayer (MGM) film productions. In fact, not only is the familiar image of *Leo the Lion* a trademark, the *sound* of the lion's roar is also a subject of trademark protection.[27] *Sound* trademark protection also subsists, for example, in Nokia mobile phones' default ring tone, Intel operating system's chime-like five-tone audio progression, the famous distinctive yell of Tarzan, and the theme tunes of the Looney Tunes cartoon series. Other real-life examples of unconventional trademarks are the unique *shape* of Coca-Cola bottles used to package its soft drink, the triangular *shape* of Toblerone chocolate bars, and the pink *color* applied to Owens-Corning fiberglass insulation products.

There are also marks which are prohibited from trademark protection. Some examples of these marks are purely descriptive marks and deceptive marks. Marks that are merely descriptive of the products or services are not registrable. For example, descriptive notions such as "beautiful" or "top" (describes the quality) and "two" (describes the quantity) can never be innately unique and hence are not protected by trademark. Moreover, marks that are of such nature that attempt to deceive or mislead the public (for instance as to the nature, quality, or origin of the wares or services) also deserve no protection under trademark law.

Registrability of Trademark

To ensure eligibility for trademark registration, the subject matter must be unique in the sense that it is capable to serve as a source indicator to identify the commercial origin of a particular product or service. Furthermore, legislation of most countries generally allows trademark protection only for marks that are visually perceptible. In other words, they require the mark to be capable of being illustrated graphically. This requirement is always an issue for some nontraditional marks, for instance, sound and scent.

Rights of the Trademark Owner

Depending on the legislation of a particular country, trademark registration may or may not be required by a government authority. The owner of a registered trademark is given the statutory right and exclusive monopoly with regard to the trademark. He or she has the sole right to use it (such as branding) to distinguish wares or services. The owner has the right also to prevent others from using the protected mark, or any other significantly similar marks. If the mark is being exploited by a third party without obtaining permission from the registered trademark owner, or when imitation arises, the owner can basically rely on his registration as evidence of his right to the mark and take legal action to sue for infringement.[28]

Conversely, if the owner of an unregistered mark encounters the same situation, he may need to resort to having the upper hand, depending on his goodwill and reputation. In many scenarios, this requirement may land the owner in a difficult situation if the use of the trademark, or the business associated with it, is relatively new or not-yet established in the industry. Therefore, it is always advisable to acquire registration for a trademark so that its usage is not easily exploited by unauthorized parties. After all, it is definitely cheaper to register your mark than painstakingly defending it if it is stolen by a competitor. The term of protection for a trademark varies among territories. However, it can be renewed indefinitely beyond the stipulated duration on the condition that additional fees are duly paid.

In most countries, the symbols normally associated with trademarks are ©, ®, and SM. Commonly pronounced as "R-in-a-circle" or "Circle-R", the ® notice can only be used for registered marks, and is formally protected under the trademark law.[29] On the other hand, the ™ and SM notices are usually used to represent unregistered trademarks and service marks, respectively, which are not governed by trademark law. They are commonly used by someone to claim rights to the marks without or prior to obtaining formal registration.[30]

Protecting Your IC-Related Invention: Which One to Use?

The wide array of inventions that can exist in the IC industry are practically limitless—from items as simple as a word or logo attached to an IC product to complex designs of IC devices incorporating millions of transistors; and from items as ordinary as household appliances to embedded systems for military applications.[31,32] With the present IP system, there are various ways we can protect our designs and intellectual efforts that give birth to IC-related innovations. In fact, the IP vehicles we have discussed in the earlier sections of this chapter, patent, copyright, layout-design of IC, industrial design, and trademark, all provide specific legal protection to various types of IC inventions in their own ways. For some IC products, multiple forms of IP can coexist concurrently to provide maximum protection to the subject matter embodied within the product. For example, memory devices such as random-access memories* (RAMs) and read-only memories† (ROMs) that are used for storing sets of instructions and information (also known as firmware‡) for microprocessors, are protected by several types of IPs—the designs of topographies embedded in the ICs are protected by the explicit form of sui generis protection, commonly known as the Layout-Designs of ICs Act (the Act); the mark engraved or attached onto the chip is protected by trademark; the sets of instructions or codes stored may be protected as literary works under copyright, and in certain circumstances they may be subject to patent protection as industrial methods.[33]

It is worth noting that the establishment of the Act as a relatively new concept to protect IC layouts is not meant as a replacement for the role of patent in IC layout protection.[34] This can be derived from the fact that the scope of protection of the Act, which does not include the patentable aspects such as ideas, concepts, methods, processes, systems, techniques, or information that may be manifested in the layout-design of an IC. In most

*A type of volatile memory, RAM retains its memory patterns as long as power is being supplied. Conversely, its contents are lost once the power is removed.

†A type of nonvolatile memory, ROM preserves its memory patterns even with the removal of power supply.

‡Firmware protection is explained later in this chapter.

countries, patent law is still the primary form of IP that is responsible for the protection of circuit design, since it provides much more extensive protection to ICs.[33,35] Therefore, there may be a coexistence of rights arising from the dual protection of patent law and the Act for the owner of a single piece of layout-design. Patent law is applicable for the protection of, among others, the elements of chip designs,[36] the combination or characterization of virtual component blocks,[37] or the structure or method of operation of electronic circuitries embodied within an IC product,[33] on the condition that they meet the requirements of novelty, inventive step, and industrial applicability. Figure 4-1 shows the types of IP that are applicable to different subject matter available at various stages of a large system-based circuit design. On top of that, the methods or processes for chip fabrication used in chip production plants may also be protected by patent law.[38]

As a result of the enactment of the sui generis type of protection, copyright in most countries no longer subsists for IC topographies protection.[34] However, the schematic diagram, which is a pictorial representation of a circuit, may still be covered under copyright protection as a literary or artistic work. In spite of that, copyright is considered a rather weak form of protection, because copyright, by its nature, only protects the expression of the schematic drawing; the information encompassed by the circuit therein is unprotected. As a result, copyright does not give protection to its holder should reverse engineering* of the final chip product occur.

IP is many times the lynchpin determining the success of an IC business. As a matter of fact, the choice of the most appropriate type of protection to be employed for an IC-related innovation is in effect a critical business decision that has to be made by its owner.[39] The owner has to fully understand the best form, or perhaps several forms, of IP to be used for protecting his invention before finalizing the decision. During the decision-making process, the owner is advised strongly to consult an attorney with relevant IP knowledge and experience in the specific field of ICs, as IP law is a complicated and ever changing domain, hence requiring someone who is intimately familiar and constantly keeping abreast of the changes to secure protections for creative works.

Computer Software Protection

This section entails a discussion of the primary forms of IP protection for computer software. More than three decades ago, with the increasing interest and popularity of the computer, there were extensive debates on the

*The discussion of reverse engineering is presented in Chapter 6.

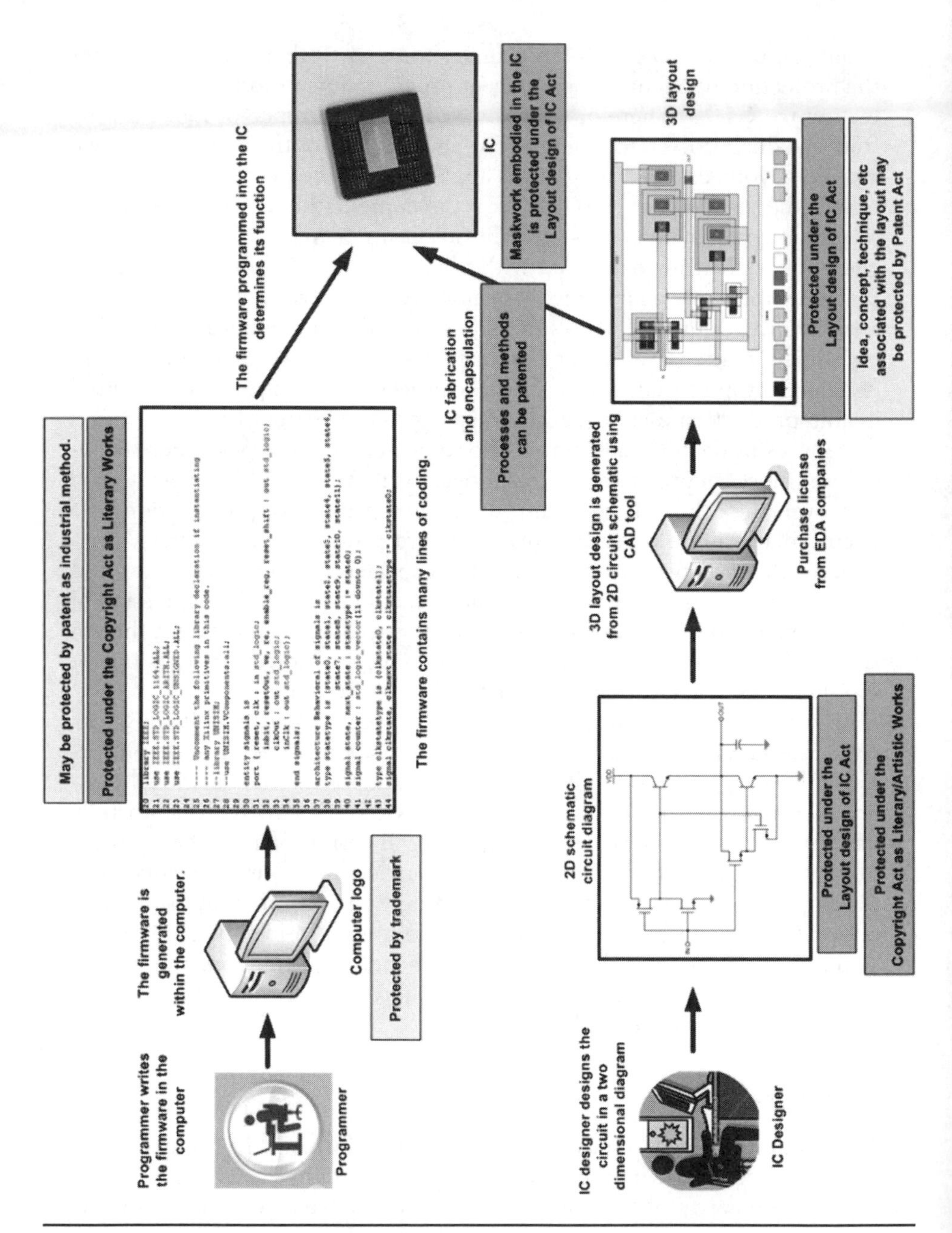

Figure 4-1 Types of IP protection for different subject matters at various stages of circuit design

most suitable form of legal protection for computer software: copyright, patent, or a sui generis regime (a special form of protection created especially for computer software). Although no firm consensus was reached, it was then generally accepted that computer software should fall under the jurisdiction of the copyright system whereas the hardware utilizing the computer software or its related inventions should be protected by the patent system.[40] The rationale behind this general acceptance hailed from the fact that the copyright and patent systems bestow different aspects of protection. While copyright law protects only the expressions but not the ideas, concepts, methods of operation as well as mathematical concepts, patent law protects an invention that gives a new and inventive way of doing something, or provides a new technical solution to address a particular problem. In fact, in the 1970s, the U.S. Supreme Court ruled that software was not patentable on the grounds that it is fundamentally a mathematical algorithm.[41]

However, with the exploding business opportunities involved in the software market, there has been a paradigm shift with respect to the most suitable mode of software protection—copyright versus patent system. As a result, the whole saga of copyright versus patent system for software protection was revisited.[3] In one of the most notable legal cases involving the question of patentability of software, *Diamond vs. Diehr,* the U.S. Supreme Court ruled in 1981 that the grounds for patent denial to an invention should not rest solely on the fact that mathematical formulae constitute one of its claims.[42] Rather, the invention should be viewed as a whole. Moreover, in 1998, the U.S. Federal Circuit Court of Appeals ruled in favor of the fact that instead of relying on categories of subject matter, the invention should be considered for its practical functionality and subsequently tested together with the other requirements of novelty and inventive step.[43] Another breakthrough took place in 1999, when it was ruled that algorithms are patentable, since they confine the application of a general-purpose computer to accomplishing specific-purpose tasks by carrying out functions as dictated by the software.[44]

Similar breakthroughs were seen in Europe. In 1979, Vicom Systems, a corporation headquartered in California, filed an application in Europe for a "method and apparatus for improved digital image processing."[45] One year prior to the application in Europe, Vicom had already filed for an equivalent application for U.S. patent in 1978, which was granted in 1982. However, interestingly, the Examining Division of the European Patent Office rejected the application on the grounds that the method claims were either related to a mathematical method or that there was no addition of a technical feature and, therefore, rendered it not patentable.[3] This position was subsequently overturned by the European Patent Office Technical Board of Appeal, rationalizing that digital image processing is not an abstract process but a *real world activity*[46] and held that "even if the idea underlying an invention may be considered to reside in a mathematical method, a claim directed to a technical process in which the method is used does not seek protection for the mathematical method as such."[47]

Ultimately, numerous court cases in both the United States and Europe have helped to determine that computer programs are, indeed, patentable.

The reader may wonder—since software can already be protected by copyright, why then should inventors fight so hard to have it protected under the patent system? The answer essentially lies in the strength of protection accorded by patent law.[48] Recall that copyright law can only prevent the copying of a particular expression of an idea but does not protect the idea itself. With respect to computer software, copyright law is effective to prevent the exact duplication of whole or part of a software program. Hence, copyright law does not protect the inventor against the creation of a rival program that is based on exactly the same idea as a preexisting program. On the other hand, a patent owner has the right to prevent all others from making, using, or selling the patented invention. In relation to computer software, an issued patent can prevent others from adopting a certain algorithm (such as the GIF image compression algorithm) without permission, or can prevent others from creating software programs that perform a function in a certain way. Consequently, patent law can accord much better benefits and rights to a software inventor than copyright law. This is obvious from the lawsuit victory of Stac Electronics against Microsoft over a data compression patent and the award of $120 million. With better understanding of the potential benefits of software patents, more patents are being issued. Based on the record of the Software Patent Institute, the number of *true software patents* issued every year runs into the thousands, covering such areas as business software, expert systems, compiling functions, operating system techniques, and editing functions.

Despite the benefits provided by patent over copyright, it does not mean that the former by itself can offer sufficient protection for software. This is because in the context of intellectual property, software is essentially a true hybrid.[49] While software manifests itself as codes and wordings, it is not purely literary—it also embraces functionality, i.e., instructions intended for performing a certain task—it is this feature that apparently sets it apart from normal writings. Hence, software is viewed as a hybrid, because it conveys both intellectual components and at the same time is capable of physically reducing them to practice by simply using a computer. In view of this special nature, the authors believe that the dual-protection strategy based on both patent and copyright laws gives software optimum protection—while patent protects the functioning implementations of concepts, copyright protects the modes of expression.

Firmware IP Protection

Many of us are often puzzled over the differences between firmware and software, because they are closely related to each other. Firmware is actually the software that is embedded in a hardware device for the purpose of controlling it. It is the low-level programming code capable of being programmed into an IC chip and is, thus, considered a permanent entity of the hardware. But it is also tightly linked to a piece of hardware and has little meaning outside of it. A firmware module is developed and tested by using code simulation the same way as software.[50] Firmware is required to keep a computer operating even with the absence or removal of all other software.[51] Firmware usually executes a basic set of instructions whereas software is a program designed to run on a computer's operating system.

A typical example of an IC chip incorporating firmware is the nonvolatile memory storage device such as ROM, programmable ROM (PROM), electrically erasable PROM (E^2PROM), or flash memory generally used to control the function of a microcontroller. Another example is the basic input/output system (BIOS) chip run by a computer to boot up the system. Additionally, firmware is also stored as software embedded in household electric appliances, such as digital versatile/video Disc (DVD) players, washing machines, microwave ovens, and refrigerators. Firmware, which appears as lines of computing coding meticulously written by a programmer, is a subject for protection under copyright as a literary work. It may also be granted patent protection.[42,49] The other forms of IP protection, such as layout-designs of ICs, industrial design, or trademark provide no defense to firmware.

What Information Can Be Extracted from an IC Patent?

Of all the various vehicles of the IP regime, patent (or utility patent as it is known in the United States) is perhaps the most regularly sought. In this section, we help you through the reading and digesting of the information contained in an issued patent so that you can extract pertinent information from the disclosure of the claimed invention without difficulty. It is easy to read and understand a patent related to IC or any other patents in a different technical area of specialty. They are all recorded in a well-established format which is broadly adopted. We have arbitrarily picked a recent U.S. utility patent related to the IC industry and use it as an illustrative example—a copy of this patent is shown in Figure 4-2 at the end of this chapter for easy reference. All U.S. patents are issued by the United States Patent and Trademark Office (USPTO).

Cover Page

A patent related to an IC typically has four sections (1) cover page, (2) drawings, (3) specifications, and (4) claims. Let us now begin with the cover page of the patent. You will notice bracketed numbers adjacent to each section on the cover page. These are the INID codes, an acronym for Internationally Agreed Numbers for the Identification of Bibliographic Data, which are standardized codes used universally to enable classification and identification of important information printed on a patent even if the language is entirely foreign to you.[52] Thus, even though the cover page of patents granted by countries around the world may have some disparities in their designs, international efforts in harmonizing the format and appearance of patents, such as the INID codes, helps the reader to understand patents. We use these codes to identify the sections for ease of discussion and comprehension.

(12) **Document Type:** The most eye-catching bold wording on the cover page is *United States Patent* printed at the top left hand corner of the page which identifies the document type. The last name of the inventor appears right below the type of document. In cases in which there are multiple inventors, only the last name of the first inventor appears, followed by *et al.*

(10) **Patent No.:** A patent number is used for the purpose of identifying a patent publication. A patent number has a two-character country code, *US*, preceding a string of numbers which clearly denotes that it is a U.S. publication with an assigned patent number. This is followed by a letter and, in most cases, a number, the combination of which is known as the *kind code* used to differentiate the kind or type of patent document. Different countries may have their own set of codes to represent varying patent types. Some examples of kind codes (and their interpretations) that commonly appear on U.S. utility patents are:[53]

A1—Patent application publication

B1—Utility patent grant, without pregrant publication

B2—Utility patent grant, with pregrant publication

These codes are only indicated on the face of patent publications beginning January 2, 2001. The patent number issued was *US 6,998,953 B2*, therefore, it is read as U.S. Patent no. 6,998,953 with existing previously published patent application publication. For official correspondence with the USPTO, the patent attorney or inventor may use the last three digits of the patent number, in this case *the 953 patent* to refer to our sample patent.[54]

(45) **Date of Patent:** The date of patent is stated beneath the patent number. This is the date on which the patent was issued by the USPTO,

which also means that the patent becomes enforceable starting from this date. One trivial yet interesting point to note is that this date nearly falls on a Tuesday for all patents granted. The normal practice of USPTO is to issue patents once a week, on Tuesdays.

(54) **Title of the Invention:** The title of the patent should be indicative or suggestive of the invention disclosed in the patent publication. As a general rule, it should be short and specific.[55,56] In our sample patent, the title is High Performance RF Inductors and Transformers Using Bonding Technique.

(75) **Inventors:** The inventor field reflects the full name of each and every individual who has played an important role in creating the invention, normally along with their place of residence. In some cases, the complete residential address (not work address) is also provided. Please note that in the United States, only individual inventors are to be listed in this field because companies or corporate entities cannot qualify as inventors. This practice is in contrast to legislation in most other countries in which companies can be named as the inventor.

(73) **Assignee:** If the ownership of the patent is assigned to another individual, a company, or any other parties other than the inventor(s), in many cases the employer of the inventor(s), the name, and location of the assignee company appears in this field. The assignee thus has the authority to exercise patent's rights and exploit the invention on behalf of the original patentee. Conversely, this field is blank if the patent is not assigned. However, the assignee information printed on the cover page may not be correct or up-to-date. For example, should the inventor(s) decide to relinquish patent rights to someone else by executing assignment at a later stage after the patent grant, this information will not be included in the patent document. Moreover, it is at the discretion of the attorney or inventor whether or not to list the assignee when filing for patent. Therefore, to secure accurate information in case such circumstances occur, you should query the latest assignment status of a particular patent prior to any decision making, and this can be done via the USPTO website.[†]

(*) **Notice:** This section indicates whether there is an extension in the patent term. In some scenarios, such as a long delay in examination or processing of the patent application, there is adjustment in the delay, and the term of patent validity is extended beyond the normal duration of 20 years. In this particular instance there was no extension of the term.

[†]Patent assignment query menu. U.S. Patent & Trademark Office. [Online]. Available: http://assignments.uspto.gov/assignments/?db=pat

(21) **Application No.:** This is the serial number of the application assigned by the patent office upon filing of the patent. It is prefixed by a two-digit code to indicate the patent classification and time span within which the patent application is filed, followed by a six-digit number that runs in a sequence according to the order in which the application is received. This method of serial number representation can vary in other countries.

(22) **Filed:** The date the application was filed appears here.

(65) **Prior Publication Data:** The patent may have been published while its application was still pending before official patent issuance. In such cases, the number and date of the application publication is entered here. In our example, the patent application was published on August 4, 2005, and was assigned a publication number *US 2005/0167828 A1*.

(60) **Related U.S. Application Data:** If the patent is legally associated to any other earlier domestic applications or patents, those should be recorded in this field. On the other hand, if the patent application has been filed in a foreign country prior to filing with the USPTO, a statement similar to **Foreign Application Priority Data,** INID code (30) should exist and entail relevant information, including the date, application number, and the country in which it has been filed.[57]

(51) **Int. Cl.** and (52) **U.S. Cl.:** To simplify the process of patent search, all patents are classified or indexed hierarchically according to their technical subject matter or field of invention. These fields identify international and domestic search classifications, specified in classification class and subclass, within which a patent is categorized. Considering the domestic classification of our current example, the patent is classified in class 336, subclass 200. Meanwhile, the international patent classification system uses both letters and numbers to symbolize and categorize the intended functions of patents.[58]

(58) **Field of Classification Search** or simply, **Field of Search:** This section provides a list of the class and subclass of classifications searched by the patent examiner. This information can provide some useful search directions for an inventor or attorney, who is conducting a patent search or drafting a patent application in an area that seems similar to the particular patent.

(56) **References Cited:** During the examination process of a patent application, the patent examiner needs to search through the prior arts before making a decision on whether the claimed invention deserves patent protection. All relevant publications considered are listed in this field. Other than those cited by the examiner, this record also includes the prior

arts listed by the inventor or attorney through an information disclosure statement submitted at the time of patent filing. The references cited here are not limited to U.S. patents, but rather encompass foreign patents and also other pertinent nonpatent literature, typically scientific journals. For our 953 patent, the first in the citation are U.S. patent documents with the patent number, date of issue, last name of the first inventor, and search classification stated. Beneath the U.S. patent is a list of foreign patent citations.

The names of the **Primary Examiner** and sometimes **Assistant Examiner** are recorded in this section.

(74) **Attorney, Agent, or Firm:** The name(s) of the patent attorney, agent, and/or patent law firm who executed prosecution of the patent application on behalf of the inventor or assignee is set forth in this field. Our patent example indicates that in this application, there are two patent attorneys representing the applicant but the law firm information is omitted.

(57) **Abstract:** This is a brief paragraph constituting a succinct description of the disclosure in less than 150 words.[59] It resembles the abstract of a scientific publication.

Following the abstract is a mere enumeration of the **number of claims** and **drawing sheets** that are contained in the patent. There are four claims and eight drawing sheets in Yeo's patent. It is also possible that some patents do not carry any drawings at all, but there must at least be one claim.

The last entry on the cover page is usually a **representative drawing** selected from the figures included in the patent document. If you flip through all eight drawing sheets (pages two to nine) in the patent example, you will notice that the drawing that was chosen for reproduction on the cover page was actually Figure 3A illustrated on page four of the patent document.

Drawings

Following the cover page, the subsequent pages set forth the drawings of a patent. Drawings must accompany a patent application if their inclusions are necessary for a more lucid comprehension of the subject matter of the invention. The truth is that drawings are essential in the majority of patent applications; a patent in the area of ICs usually includes several sheets of drawings—so do patents involving mechanical apparatus. As the saying goes, "a picture is worth a thousand words." Used together with the written description of the invention, the drawings help to simplify the text and allow the audience to understand the patented subject matter more easily. These drawings are respectively labeled beginning with *Figure 1*. If it is required to show illustrations of previously published inventions, these figures are inserted in this section despite the fact that they are not considered part of the patented invention—they are identified as *Prior Art*. The interested reader may like to search for a U.S. Patent with patent number US 6,277,710 B1 to have a glimpse of what *Prior Art* drawings look like. In some cases, a drawing is split into a few subdrawings to show different views of a structure. By convention, the way to name such figures is by appending letters to the label numbers, e.g., Figure 1A, Figure 1B, and so on. Our current example of the 953 patent comprises such drawings. In addition, the figures in an IC-related patent generally include symbols represented in letters and/or numerals to indicate a certain segment of the invention. The usage or explanation of these symbols is covered in the specification section which follows. Note: you can search for patents from the patent searching databases readily available on the internet, some of which are listed in the reference list for your convenience.[60–64]

Creating an up-to-the-mark patent drawing can be a challenging task to novices and most inventors. Patent drawings must conform to a set of rules and guidelines specified by the patent office, such as those downloadable from the USPTO website.[65] These standard guidelines include the scale of drawings, the types of drawings, the size and type of papers used, and other information for preparing a qualified patent drawing. Therefore, while it is possible for inventors to prepare their own drawing for patent application, it is always advisable to hire a professionally trained patent draftsperson to prepare formal patent drawings of the best quality.[66] Although it is somewhat uncommon, there are indeed some patents that do not require any drawings, especially in the pharmaceutical, chemical, and biotechnology fields.[57] For example, a patent disclosing the composition of a new type of chemical or material might not need any drawings to aid in the description of the invention.

Specification

The written part of a patent disclosure, the specification, which is considered the main body of the patent document follows the drawing sheets. The specification is presented in a two-column format with the columns numbered correspondingly on each page. The elements constituting the specification of a patent should appear in an order as outlined below:

Title of the invention: This is simply a repetition of the title of the disclosed invention that appears on the cover page. The title can be as concise as having just a two-word notation, such as *Transaction Card* (US 7,377,443 B2), or as general as *Semiconductor Integrated Circuit Device* (US 7,379,319 B2). It can also be as detailed as *System and method for determining and/or* *transmitting and/or establishing communication with a mobile device user for providing, for example, concessions, tournaments, competitions, matching, real-locating, upgrading, selling tickets, other event admittance means, goods and/ or services* (US 7,386,517 B1). Whatever it is, the rule that generally applies is to keep the title as brief and specific as possible without exceeding 500 characters in length. Attention, please! The length is based on character count, not word count.[56]

Cross references to related applications: This section states the earlier patent applications that are related to the present application. It normally contains duplication of information on the cover page, under Related Application Data. The present status of the related patent applications are normally provided here—either the application was approved (in this case, the patent number will be given) or abandoned.

Background of the invention: There are two parts that contribute to this section (1) the Field of the Invention and (2) the Description of the Prior Art. The objective of this section is to provide the audience with a general idea regarding the scientific field of the particular invention and to describe the previous work done or published in order to point out why there is a need for the intended invention. The two parts forming this section are discussed in greater detail as follows:

(1) **Field of the invention:** The key technological area to which the invention pertains is presented in this field. It typically consists of two segments. The first segment is a general statement that gives a broad description of the field of invention and is followed by the second segment of a more specific statement conveying additional information of the scientific realm into which the patent falls. This is illustrated in our 953 exemplary patent.

(2) **Description of the Prior Art:** This section relates a description of selected prior art that the patentee feels most relevant to the claimed invention and often appears in narrative form. The prior arts may include patents, journals, or other technical articles published prior to this patent document. The preceding work may also be explicitly listed, as shown in Yeo's patent. In most cases, the drawbacks associated with the prior art are identified and stated in this section. Most importantly, the potential capability of the current invention to overcome any problems should also be emphasized. The idea is to relate what actually motivated the current invention.

Summary of the invention: This includes a concise summary of the nature and substance of the invention that should commensurate with the invention as claimed.[67] In other words, it should point directly to that precise invention in question, subtracting stereotyped statements of mere generalities, that would apply equally well to other prior patents—that is, what sets it apart from the abstract of the disclosure. The summary is usually drafted in nontechnical or layman terms, as it is typically intended for a jury audience with little knowledge in the technical area of the subject matter.[68] The objects or results attainable by the claimed invention, the advantages of using the invention, the way it contributes to addressing the problem encountered by existing arts, and the utility of the invention may all be stipulated here.

Brief description of the drawings: In our patent example, the brief drawing description is presented in column three, line 14 of the specification. This section is designed to specify what each figure in the drawing sheets shows in a succinct manner, typically one sentence for each figure.

Description of the preferred embodiment: This is an illustrative and detailed description of the best (preferred) way the invention can be implemented or embodied. It is generally the most lengthy section of a patent document. It is the obligation of the patentee to disclose his or her invention in its entirety in exchange for 20 years (normally) of monopoly rights. It serves to impart new knowledge to the public, teaching them exactly how to make use of the invention for which a patent is solicited. In the United States, the obligation to disclose an invention in a patent is stated, in part, that: the specification shall contain a written description of the invention, and the manner and process of making and using it, in such full, clear, concise, and exact terms as to enable any person skilled in the art to which it relates, or with which is most nearly connected, to make and use the same, and shall set forth the best mode contemplated by the inventor of carrying out the invention.[69]

In essence, this section provides an in-depth description of the invention and discloses the exact method of operating the invention. Any person of ordinary skill in the art of the invention should be able to make and use this particular invention. Furthermore, by reading this section, one should become educated about the best mode of carrying out the invention known to the inventor at the time of the patent application being filed. It cannot be a deliberately skimpy illustration of the procedure, nor can it be a not-as-good or close-to-the-best procedure. The patentee risks having his or her patent revoked if this requirement is not fulfilled. The drawings (if any) are described in detail in this section, along with the symbols (letters or numerals) used to refer to certain portions of the drawings. Other than drawings, examples are frequently used to demonstrate the claimed invention.

Claims

Whereas a clear and complete description of the invention is crucial to enable an ordinarily skilled person to comprehend and practice the patented creation, it is equally important to apprise the public, particularly potential competitors or alleged infringers, about the legal limits or scope of the invention that is protected by the grant of patent. The section of the application that contains this information is the claims, which is technically regarded as part of the specification. In simple terms, the claims define what is covered or is not covered by the patent. We devote a separate section on this topic, because the drafting of this section is commonly perceived as one of the most critical and difficult tasks.

While an uninvited intrusion into someone's house is called trespass, invasion into the territory of an inventor's patent claims is called infringement. The claims are extremely crucial, because they are the ultimate deciding factor of the success of a patented invention. If they are poorly drafted, even an exceptionally groundbreaking new patent will eventually fail once it is placed in the highly competitive commercial market. This is because only the claims are enforceable.[55] They are of utmost importance should there be disputes or litigation concerning unwarranted claim infringement. In conjunction with this, it is important to highlight that it is the claims, instead of the invention detailed description in the specification, that are responsible in establishing the validity of an infringement. Hence, the claims must be considered meticulously amidst the preparation and prosecution of any patent application. The claims are written in a legalistic manner that is, at times, difficult to understand.

Generally speaking, it is a good practice to claim a broader area than what is explicitly defined by the patent. The claims of an invention are normally preceded by a simple notion *What is claimed is, I (or We) claim,* or a similar phrase. This is followed by one or more sequentially numbered *paragraphs* or *items*—each denoting one claim, thus determining how many claims are stipulated in the patent. Please be reminded that there must be at least one claim in a patent. There are basically two types of claims, independent claims and dependent claims. As its name implies, an independent claim stands alone, hence is not associated with any other preceding claims. It is normally designed with considerably broad terms. The first claim is essentially an independent claim, whereas the rest may be independent or dependent.[55] It is also usually a norm for Claim 1 to be the broadest among all the claims. On the other hand, a dependent claim relies upon other cited claims (may be one or more of the independent or dependent claims), and is considered a shortcut route of assimilating every element of the prior claim(s) into the present claim by reference.[54] For instance, if Claim 2 is reliant on Claim 1, all the elements of Claim 1 are thus assimilated in Claim 2 by reference. This is on top of the elements presented in Claim 2. In this case, Claim 1 is also known as the parent or base claim of Claim 2. A dependent claim is, by nature, more narrow than the independent claim that it refers to and is used to modify and refine its parent claim.

A claim, independent or dependent, is comprised of the three segments outlined:

a) preamble
b) transitional phrase or word
c) body of the claim

The first segment of a claim, the preamble, is an introductory phrase of the claim that usually gives the name or the classification of the object being claimed. In other words, the preamble defines the working environment of the claimed invention. The second segment, which is the transitional phrase or word, might, on the surface, seem to be merely a connecting tool to link up the preamble and the body. However, it affects the scope of coverage of the claim in a patent to a large extent.[54] Let me elaborate. Some commonly used transitional terms are *comprising, consisting of, consisting essentially of,* and *including.* Contrary to their normal usage in English, these words, when used in patent claims, carry specific legal interpretations. When the word *comprising* is used in a claim, it means *not limited to* or having *at least* the elements enumerated. It is of an open-ended nature and may embrace other unnamed elements in addition to the elements listed in the claim. The word *including* bears the same meaning as *comprising* as used in a claim. Conversely, the transitions *consisting of* and *consisting essentially of* convey a less liberal scope of coverage of the claims. They are of a close-ended nature hence more restrictive. *Consisting of* denotes that *only* the elements cited can be used and any other unnamed elements are stringently excluded. The absolute strictness of *consisting of* can be mellowed down to a certain extent by replacing it with the term *consisting essentially of,* which is literally an intermediate between *comprising* and *consisting of.* This phrase is used to encompass additional elements only if they do not affect the intended properties or characteristics of the claim in any way. From the discussion above, it is apparent that the proper use of transitional phrase(s) is imperative, because it may lead to different coverage of the scope of patent. The third segment, the body of the claim, sets out the vital elements or particular steps or components or clauses of the patentable invention. The body may be constituted by subparagraphs representing multiple elements. Each element is separated by a semicolon (;) and only the last semicolon is appended with an *and.*

A product is said to have infringed upon your patent if it falls within the scope of only one of your patent claims. If that particular claim has multiple elements, the alleged infringing object must incorporate each and every element of the claim in order for infringement to occur. Infringement does not occur even if a single element of the claim is absent.

Let us consider our exemplary patent again. The section on the claims appears in column six, line nine of the last page of the patent document. There are four claims in the 953 patent, with Claim 1 being independent and all other claims depending on Claim 1. For further illustration, Claim 1 is stated this way:

> A method of fabricating a transformer in the fabrication of integrated circuits comprising:
>
> providing bonding pads over a semiconductor substrate;
>
> providing wedge bond wire underlying and connecting said bonding pads;
>
> making first input/output connections with two of said bonding pads;
>
> making second input/output connections with another two of said bonding pads; and
>
> forming a plurality of bond wire loops, each loop connecting each two of said bonding pads wherein said bond wire loops connecting to said first input/output connections form a first inductor and wherein said bond wire loops connecting to said second input/output connections form a second inductor and wherein said first and second inductors together form said transformer.

In this claim, the preamble is given by "A method of fabricating a transformer in the fabrication of integrated circuits," the transitional word used is *comprising,* and the body of the claim is formed by a list of five elements. This is an independent claim.

Claim 2 reads as follows:

> The method according to claim 1 wherein said bond wire loops comprise copper or gold.

Claim 2 is a dependent claim, which incorporates all contents of claim 1 and the new elements in claim 2.

To recapitulate, inventors or patent attorneys must exercise vigilance when writing the claims of patents. What is protected or not protected by the patent grant must be adequately delineated in the claims section. The claims should be able to preempt a reasonably wide scope of possible infringing conduct by potential rivals at various stages of the marketing scene including production, distribution, and consumer levels.[54] This is particularly important in the electronic—computer hardware, and software industries, all of which are related to ICs—due to the fiercely competitive nature of such industries.

Summary

This chapter provides full coverage of the classifications of intellectual property (IP) that are available in the present-day legal system for the protection of integrated circuits (ICs). Explicitly, the IP vehicles for IC protection are patent, copyright, layout-designs of ICs, industrial design, and trademark. These IPs are normally territorial in the sense that their protections are confined within the geographical boundary within which a particular IP is conferred. We have adopted a more generalized and open-minded approach while discussing the various forms of IP rather than specifically scrutinizing the legislation of a particular nation.

Patent law protects useful inventions which are novel, inventive, and involve industrial applicability. Copyright covers the aesthetic or artistic expression of an idea, which include literary, musical, dramatic, artistic, and other intellectual creations, whereas the layout-designs of ICs Act is a copyright-like IP vehicle that provides sui generis protection to the layout-design of an IC against unauthorized commercial exploitation. Whereas industrial design law provides protection for the outward appearance of an intrinsically utilitarian article of manufacture, trademark protection is afforded to unique signs used in conjunction with goods or services to indicate their source or origin.

Furthermore, although the various types of IP are theoretically dissimilar (except copyright and layout-designs of ICs) to a certain extent, more than one type of them may be embodied in the same article or subject matter. In view of that, a separate section has been devoted to discussing the forms of IP that are most suitable for protecting IC-related inventions. Furthermore, a study on the prominent types of IP vehicles for software-related legal protection suggests that this category of invention is generally subject to either copyright or patent protection, or both, depending on national laws and regulations. However, it has also been reported that there is a universal trend in favor of adopting patent protection for computer software over that of copyright. We have also dedicated a section providing an easy-to-read guide to interpreting and extracting information from an IC patent.

(12) **United States Patent**
Yeo et al.

(10) **Patent No.: US 6,998,953 B2**
(45) **Date of Patent: Feb. 14, 2006**

(54) **HIGH PERFORMANCE RF INDUCTORS AND TRANSFORMERS USING BONDING TECHNIQUE**

(75) Inventors: **Kiat Seng Yeo**, Singapore (SG); **Hai Peng Ian**, Singapore (SG); **Jiangud Ma**, Singapore (SG); **Manh Anh Do**, Singapore (SG); **Johnny Kok Wai Chew**, Singapore (SG)

(73) Assignee: **Chartered Semiconductor Manufacturing Ltd.**, Singapore (SG)

(*) Notice: Subject to any disclaimer, the term of this patent is extended or adjusted under 35 U.S.C. 154(b) by 0 days.

(21) Appl. No.: **11/095,268**

(22) Filed: **Mar. 31, 2005**

(65) **Prior Publication Data**

US 2005/0167828 A1 Aug. 4, 2005

Related U.S. Application Data

(60) Continuation of application No. 10/448,882, filed on May 29, 2003, which is a division of application No. 09/556,423, filed on Apr. 24, 2000, now Pat. No. 6,586,309.

(51) **Int. Cl.**
H01F 5/00 (2006.01)

(52) **U.S. Cl.** **336/200**; 438/381

(58) **Field of Classification Search** 336/83, 336/192, 200, 232; 257/531; 438/329, 381
See application file for complete search history.

(56) **References Cited**

U.S. PATENT DOCUMENTS

3,614,554 A *	10/1971	Shield et al.	....	257/531
4,103,267 A *	7/1978	Olschewski	....	336/65
4,777,465 A *	10/1988	Meinel	....	336/65
5,519,233 A *	5/1996	Fukasawa	....	257/275
5,543,773 A *	8/1996	Evans et al.	....	336/183
5,640,127 A	6/1997	Metz	....	330/298
5,767,563 A *	6/1998	Imam et al.	....	257/531
5,886,393 A *	3/1999	Merrill et al.	....	257/531
5,905,418 A	5/1999	Ehara et al.	....	333/193
5,945,880 A	8/1999	Souetinov	....	330/311
5,963,110 A	10/1999	Ihara et al.	....	333/28 R
6,194,774 B1	2/2001	Cheon	....	257/531
6,586,309 B1	7/2003	Yeo et al.	....	438/381

FOREIGN PATENT DOCUMENTS

EP	0883183 A	5/1998
JP	10289921 A	10/1998
WO	WO 00/10179	2/2000

* cited by examiner

Primary Examiner—Tuyen T Nguyen
(74) *Attorney, Agent, or Firm*—George O. Saile; Rosemary L. S. Pike

(57) **ABSTRACT**

A method of fabricating an inductor using bonding techniques in the manufacture of integrated circuits is described. Bonding pads are provided over a semiconductor substrate. Input/output connections are made to at least two of the bonding pads. A plurality of wire bond loops are made between each two of the bonding pads wherein the plurality of wire bond loops forms the inductor.

4 Claims, 8 Drawing Sheets

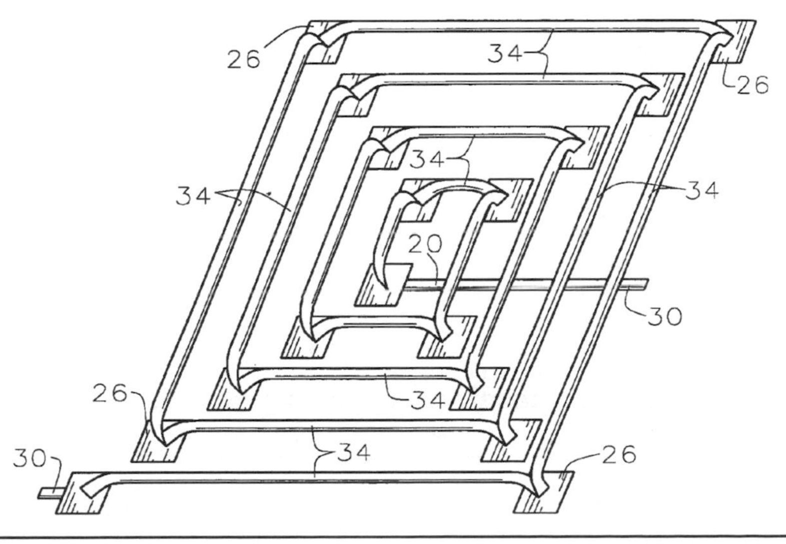

Figure 4-2 U.S. patent 6,998,953.

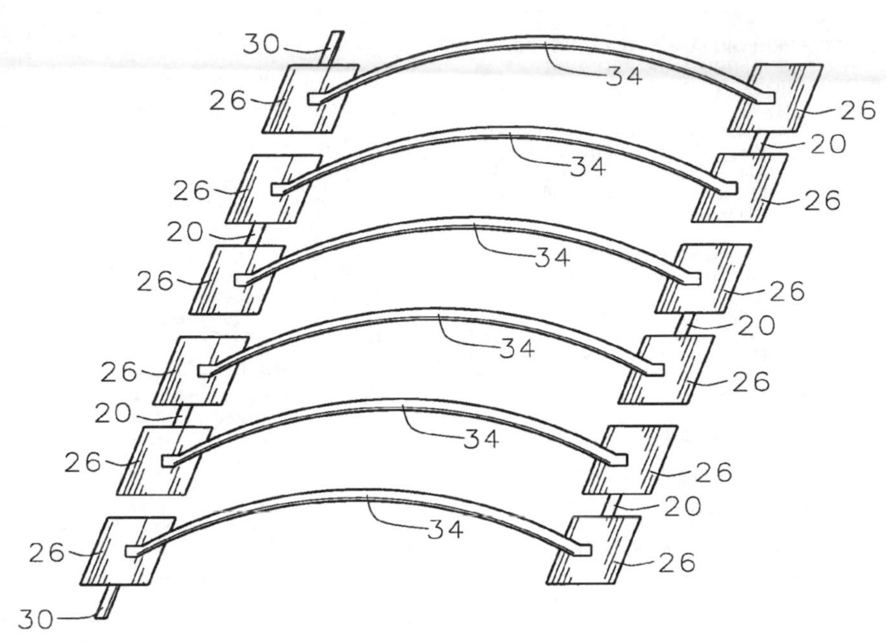
30
26
34
26
20
26
34
26
20
26
34
26
26
34
26
20
26
34
26
20
26
34
26
20
26
34
26
30

FIG. 1A

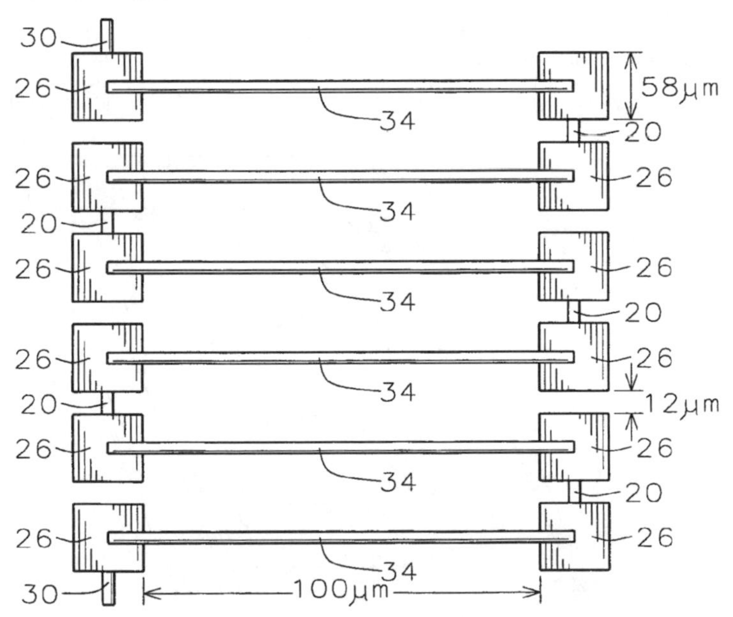
30
26
34
58μm
20
26
34
26
20
26
34
26
20
26
34
26
12μm
20
26
34
26
20
26
34
26
30
100μm

FIG. 1B

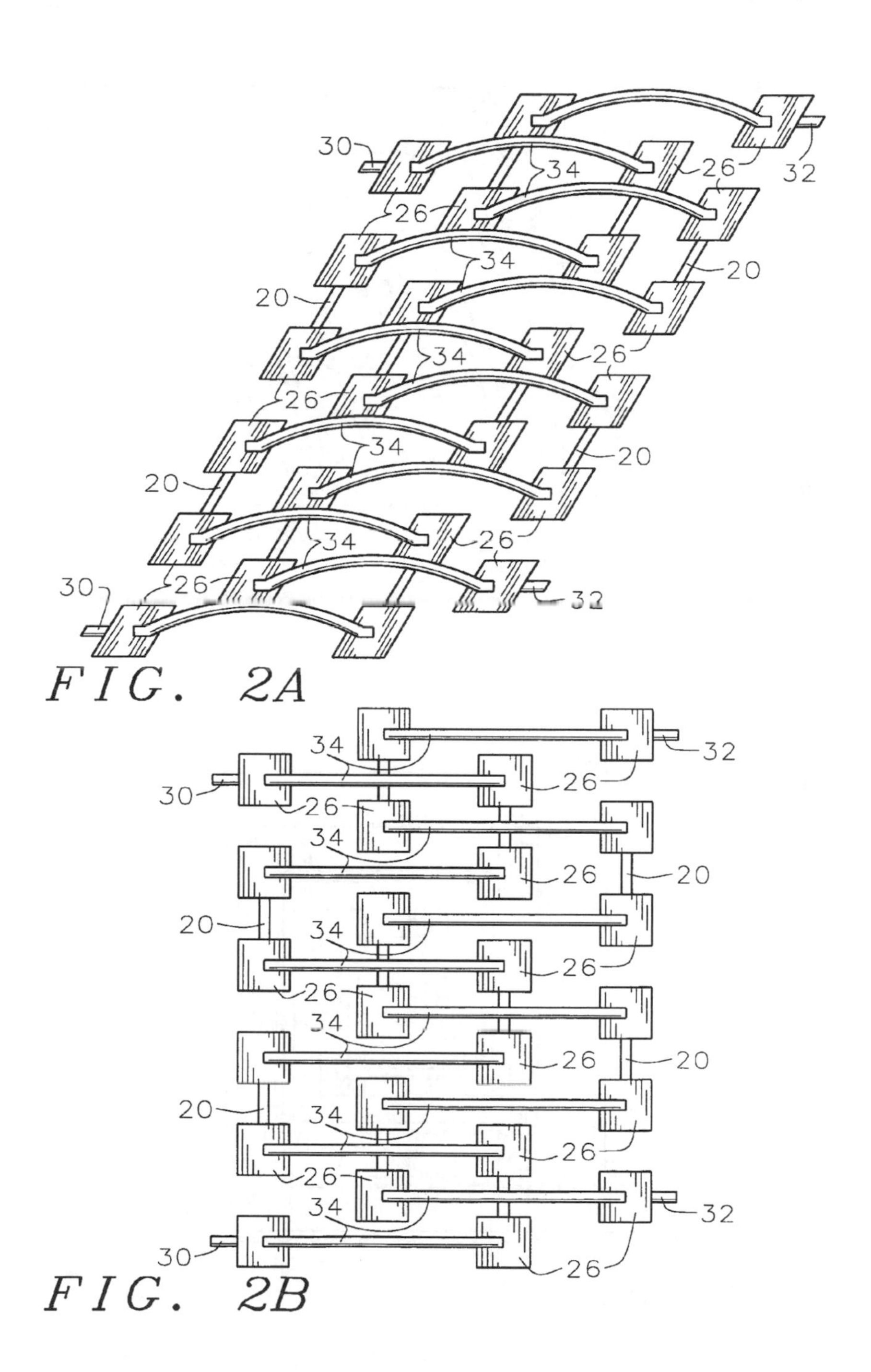
30
34
26
32
26
26
34
20
20
34
26
26
34
20
20
34
26
30
26
32
34
32
30
26
26
34
26
20
20
34
26
26
34
26
20
20
34
26
26
34
32
30
26

FIG. 2A

FIG. 2B

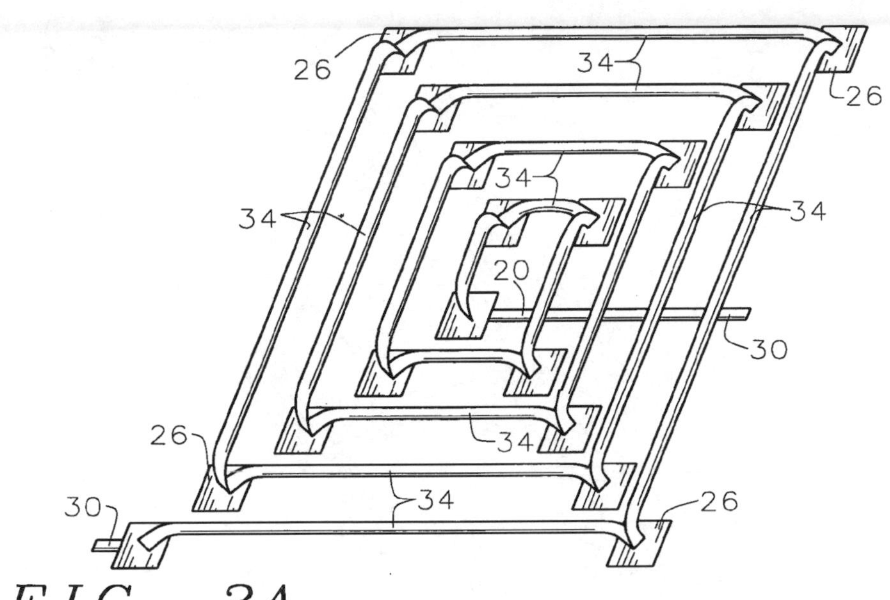

FIG. 3A

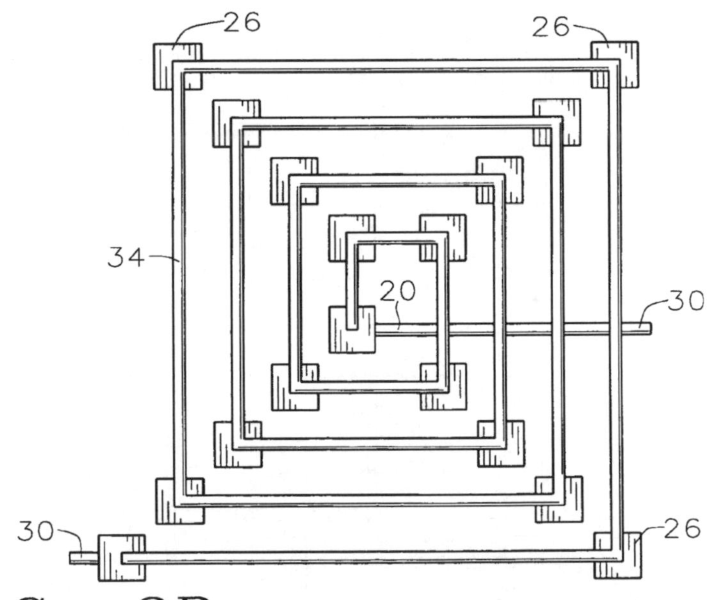

FIG. 3B

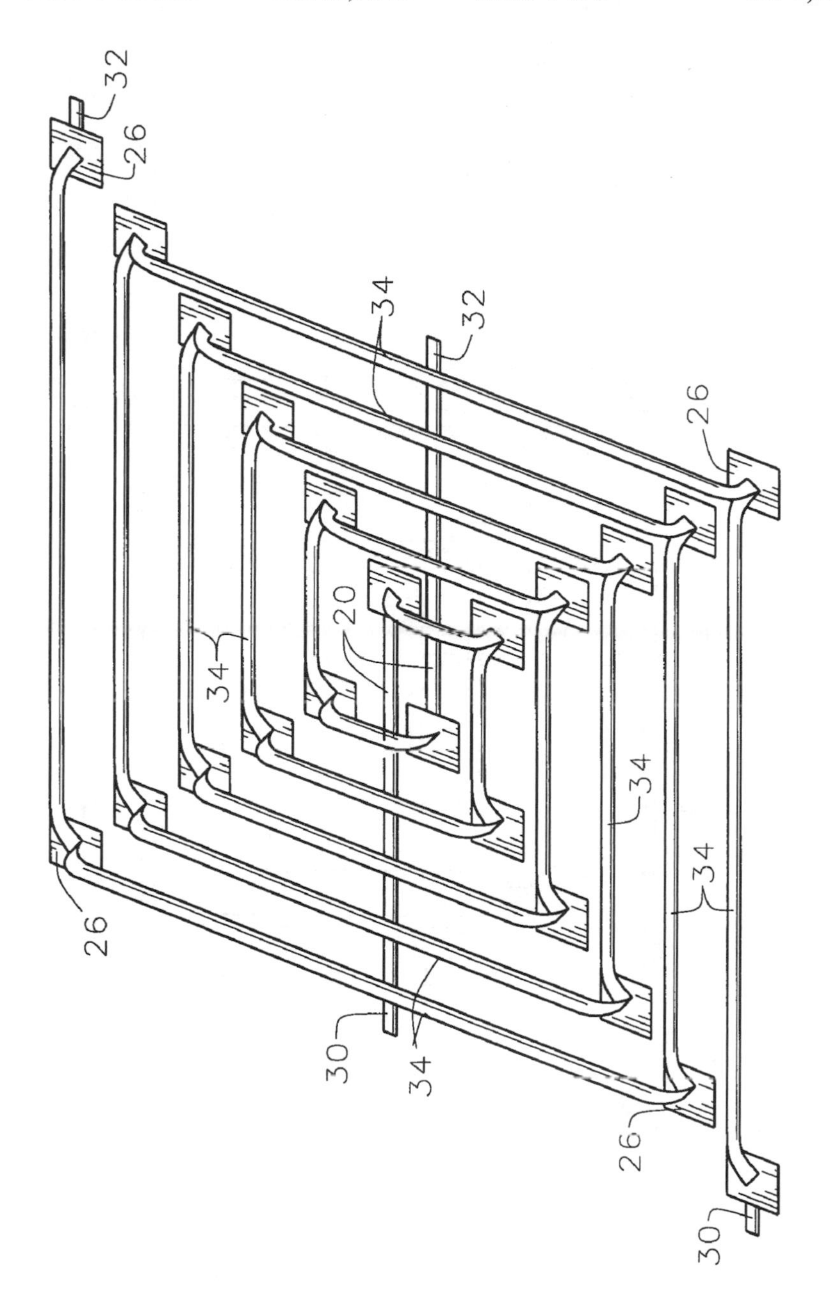

FIG. 4A

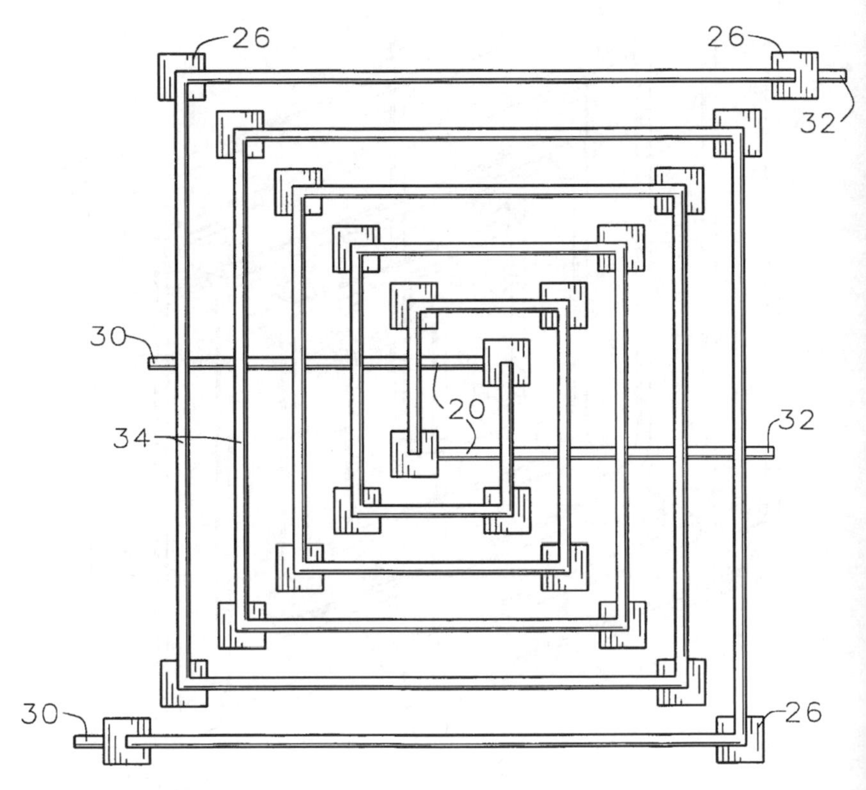

FIG 4B

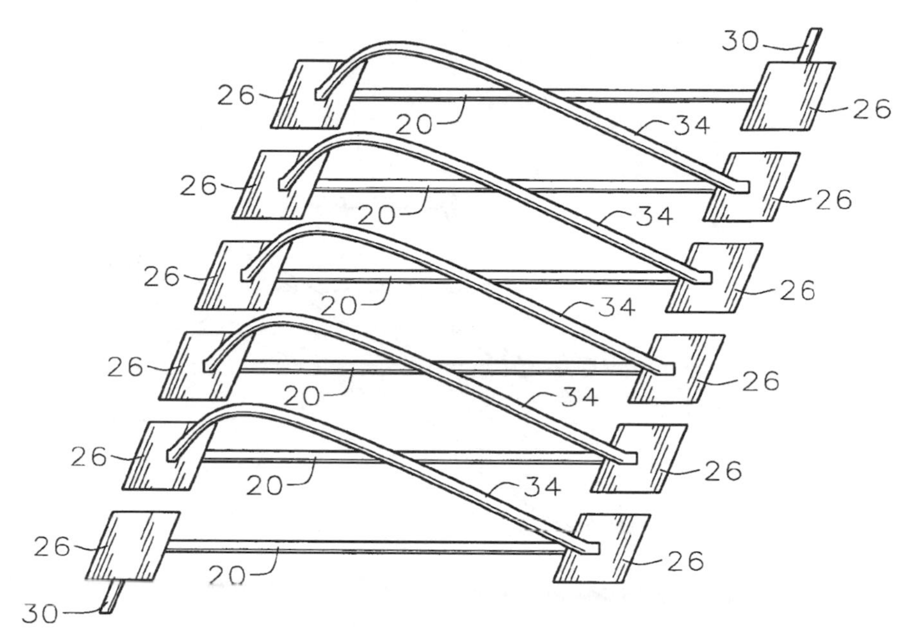
30
26
26
20
34
26
26
20
34
26
26
20
34
26
26
20
34
26
26
20
34
26
26
20
26
30

FIG. 5A

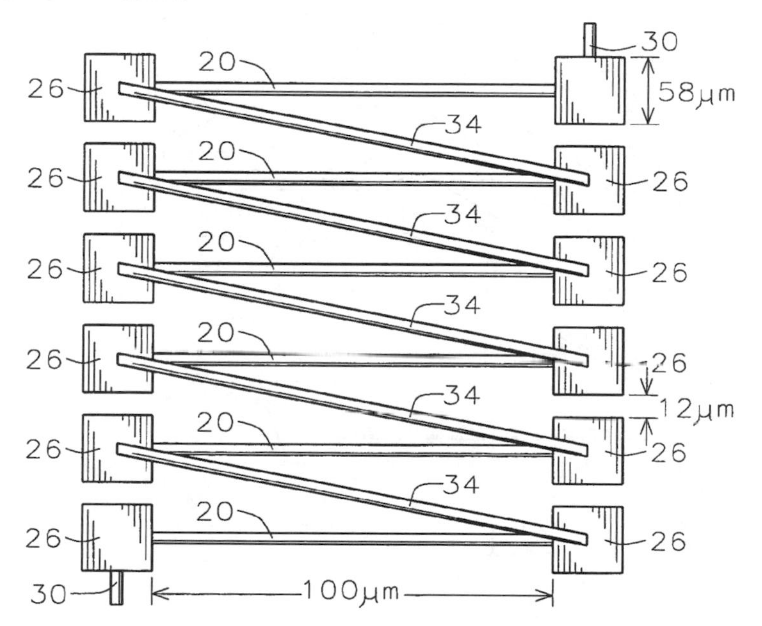
20
30
26
58μm
34
20
26
26
34
20
26
26
34
20
26
26
12μm
34
20
26
26
34
20
26
26
30
100μm

FIG. 5B

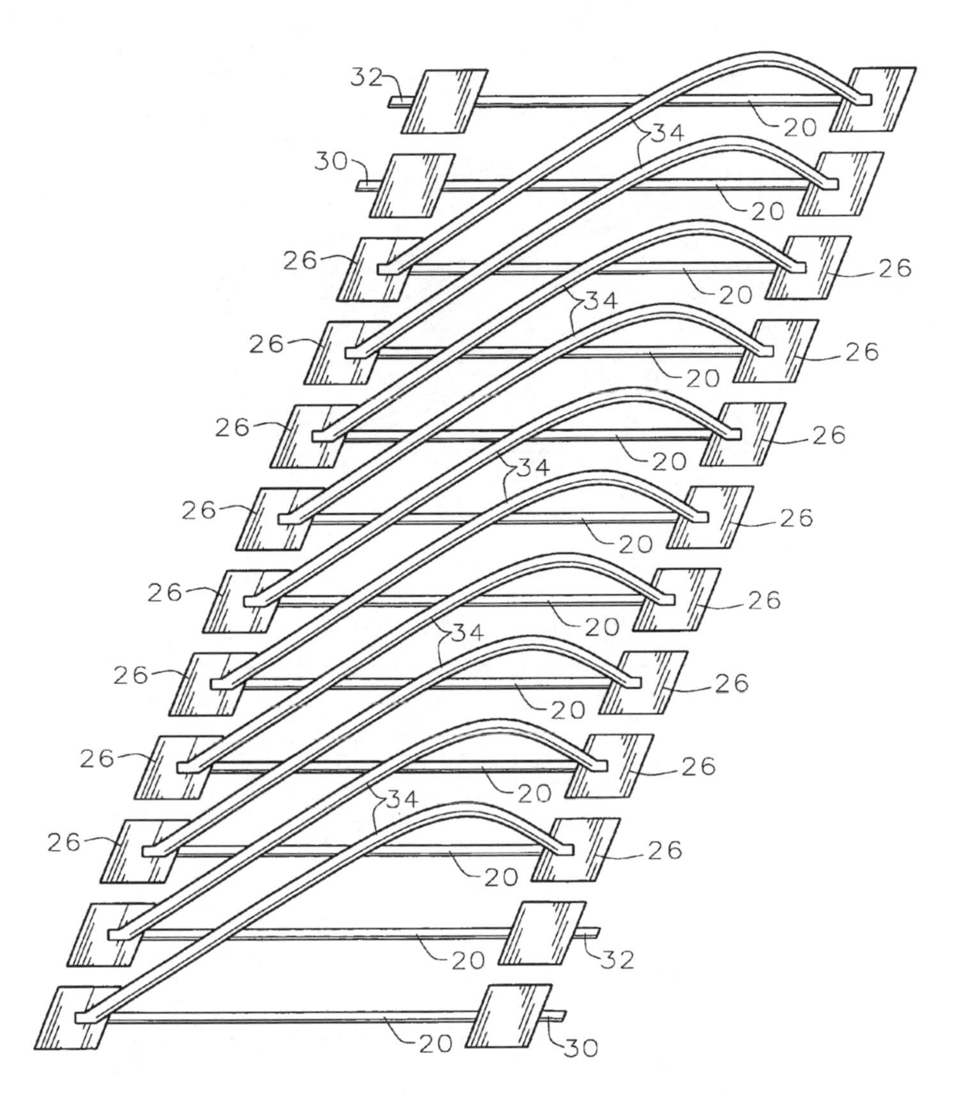
32
20
34
30
20
26
20
26
34
26
20
26
26
20
26
26
34
26
20
26
26
20
26
26
34
26
20
26
26
20
26
26
34
26
20
26
20
32
20
30

FIG. 6A

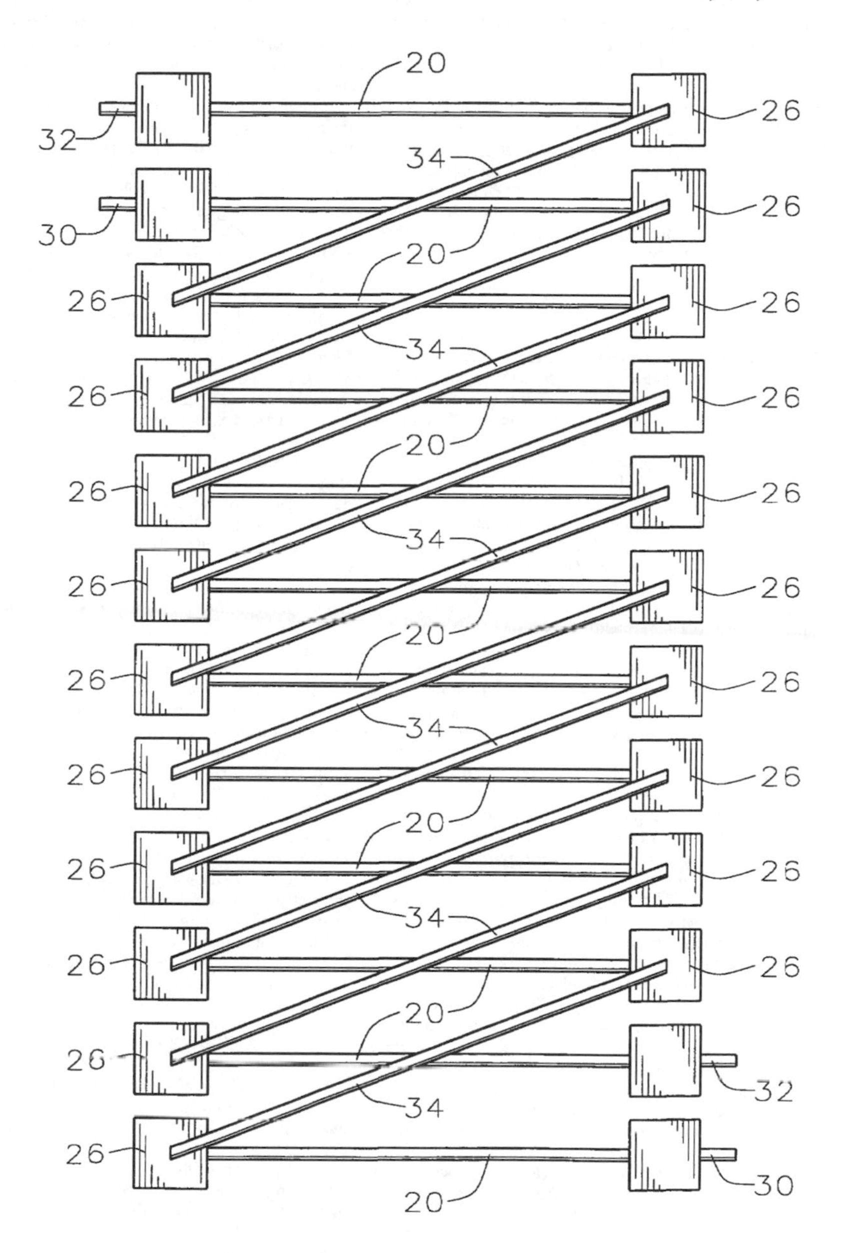

FIG. 6B

HIGH PERFORMANCE RF INDUCTORS AND TRANSFORMERS USING BONDING TECHNIQUE

This is a Continuation of U.S. patent application Ser. No. 10/448,882, filing date May 29, 2003, High Performance RF Inductors and Transformers using Bonding Technique, which is a Divisional application of U.S. patent application Ser. No. 09/556,423, filing date Apr. 24, 2000, now Issued as U.S. Pat. No. 6,586,309, assigned to the same assignee as the present invention, all of which are herein incorporated by reference in their entirety.

BACKGROUND OF THE INVENTION

(1) Field of the Invention

The invention relates to a method of forming an inductor in the fabrication of integrated circuits, and more particularly, to a method of forming a high quality inductor using bonding techniques in the manufacture of integrated circuits.

(2) Description of the Prior Art

Increasing demands for wireless communications motivate a growing interest in low-cost, compact monolithic personal communication transceivers. High performance radio frequency (RF) inductors are the key components for implementing critical building blocks such as low-noise RF voltage-controlled oscillators (VCOs), low-loss impedance matching networks, passive filters, low-noise amplifiers and inductive loads for power amplifiers, etc. Critical parameters include inductance value, quality factor, and self-resonant frequency. However, the difficulty of realizing high quality factor (Q) inductors remains a challenge especially on silicon radio frequency (RF) integrated circuit (IC) applications. Conventional inductors built on silicon have strictly planar structures and using conventional fabrication processes suffers from several limitations. Most structures and methods currently used for fabricating high Q inductors are in hybrid circuits, monolithic microwave integrated circuits (MMICs), or discrete applications which are not readily compatible with silicon VLSI processing. Consequently, the ability of integrating high quality factor (high Q) inductors on active silicon is limited.

In the past, many fabricating techniques, methods, and processes were proposed to improve the performance of the integrated conductor. In fact, most of these techniques are not cost effective or practical, requiring process changes such as depositing thicker metal/dielectric layers or starting with high resistivity substrates. Expensive processes such as the selective removal of the silicon substrate underneath the inductors has been introduced to eliminate the substrate parasitic effects. However, this processing technique raises reliability issues like packaging yield and long-term mechanical stability. Also, the typical aluminum-copper (AlCu) interconnects which are found in the conventional silicon process have higher resistivity than gold (Au) metallization used in GaAs technology. This is a concern for obtaining high inductance (L) value. Another approach is to make an active inductive component which simulates the electrical properties of an inductor by active circuitry. It is possible to achieve very high Q-factor and inductance in a relatively small size this way. However, this approach may suffer from high power consumption and high noise levels that are not acceptable for high frequency applications.

Most of these complex processes used seem promising, but they are uncommon to most semiconductor processes and they will result in high production costs. Currently, the conventional spiral inductor is still the most commonly

used. This spiral inductor which is built horizontally on the substrate surface not only occupies large chip area, but also suffers from low Q-factor and high parasitic effects due to substrate losses.

A number of papers have discussed the use of inductors for new device technologies such as RF devices. The first two references listed provide more general discussions of inductors while the remaining papers discuss spiral inductors. Refer to: (1) "New Development Trends for Silicon RF Device Technologies," by N. Camilleri et al, IEEE Microwave and Millimeter-Wave Monolithic Circuits Symposium, p. 5–8, 1994; (2) "Future Directions in Silicon IC's for RF Personal Communications," by P. R. Gray et al, Proc. Custom Integrated Circuits Conference, p. 83–89, 1995; (3) "High Q CMOS-Compatible Microwave Inductors Using Double-Metal Interconnection Silicon Technology," IEEE Microwave and Guided Wave Letters, Vol. 7, No. 2, p. 45–47, February 1997; (4) "RF Circuit Design Aspects of Spiral Inductors on Silicon," by J. N. Burghartz et al, IEEE Journal of Solid-State Circuits, Vol. 33, No. 12, p. 246–247, December 1998; (5) "Novel Substrate Contact Structure for High-Q Silicon Integrated Spiral Inductors," by J. N. Burghartz et al, Tech. Dig. Int. Electron Devices Meeting, p. 55–58, 1997; (6) "Microwave Inductors and Capacitors in Standard Multilevel Interconnect Silicon Technology," by J. N. Burghartz et al, IEEE Transactions on Microwave Theory and Techniques, Vol. 44, No. 1, p. 100–104, January 1996; (7) "The Modeling, Characterization and Design of Monolithic Inductors for Silicon RF IC's", by J. R. Long et al., IEEE Journal of Solid-State Circuits, Vol. 32, No. 3, p. 357–368, March 1997; (8) "Multilevel Spiral Inductors Using VLSI Interconnect Technology," by J. N. Burghartz et al, IEEE Electron Device Letters, Vol. 17, No. 9, p. 428–430, September 1996; (9) "Si IC-Compatible Inductors and LC Passive Filters," by N. M. Nguyen et al, IEEE Journal of Solid-State Circuits, Vol. 25, No. 4, p. 1028–1031, August 1990; (10) "Experimental Study on Spiral Inductors," by S. Chaki et al, IEEE Microwave Symp. Dig. MTT-S, p. 753–756, 1995.

U.S. Pat. No. 5,886,393 to Merrill et al teaches forming an inductor in a package form using bonding wire segments. The inductor can have any shape. U.S. Pat. No. 5,963,110 to Ihara et al shows an inductor formed on a ceramic substrate using bonding wire divided into sections and having daisy-chained connections across landing pads. U.S. Pat. Nos. 5,905,418 to Ehara et al, 5,945,880 to Souetinov and 5,640,127 to Metz disclose inductors formed in whole or in part of bonding wire.

SUMMARY OF THE INVENTION

A principal object of the present invention is to provide an effective and very manufacturable method of forming a high quality inductor in the fabrication of integrated circuit devices.

Another object of the invention is to provide a method of fabricating an inductor using bonding techniques in the manufacture of integrated circuits.

A further object of the invention is to fabricate using bonding techniques a high quality inductor that operates at high frequency.

Yet another object is to fabricate using bonding techniques a high quality inductor that operates at high frequency wherein this fabrication is suitable for VLSI integration at low cost.

In accordance with the objects of this invention a method of fabricating an inductor using bonding techniques in the manufacture of integrated circuits is achieved. Bonding pads are provided over a semiconductor substrate. Input/output connections are made to at least two of the bonding pads. A plurality of wire bond loops are made between each two of the bonding pads wherein the plurality of wire bond loops forms the inductor.

Also in accordance with the objects of the invention, an inductor comprising wire bonds is achieved. Bonding pads are formed on a semiconductor substrate. The inductor comprises a plurality of wire bond loops connecting each two of the bonding pads. Input/output connections are made to at least two of the bonding pads.

BRIEF DESCRIPTION OF THE DRAWINGS

In the accompanying drawings forming a material part of this description, there is shown:

FIGS. **1***a* and **1***b* schematically illustrate in side view and in plan view, respectively, a preferred embodiment of a meander-shaped inductor of the present invention.

FIGS. **2***a* and **2***b* schematically illustrate in side view and in plan view, respectively, a preferred embodiment of a meander-shaped transformer of the present invention.

FIGS. **3***a* and **3***b* schematically illustrate in side view and in plan view, respectively, a preferred embodiment of a spiral inductor of the present invention.

FIGS. **4***a* and **4***b* schematically illustrate in side view and in plan view, respectively, a preferred embodiment of a spiral transformer of the present invention.

FIGS. **5***a* and **5***b* schematically illustrate in side view and in plan view, respectively, a preferred embodiment of a solenoid inductor of the present invention.

FIGS. **6***a* and **6***b* schematically illustrate in side view and in plan view, respectively, a preferred embodiment of a solenoid transformer of the present invention.

DESCRIPTION OF THE PREFERRED EMBODIMENTS

Existing integrated inductors have low quality factors and low inductance values and they occupy large area spaces. The proposed integrated inductor using bonding techniques, as illustrated in FIGS. **1**, **3**, and **5**, is able to overcome these problems with the use of current semiconductor IC processing.

Currently, there are three types of bonding methods that are used for high-density VLSI packaging: the thermocompression, thermosonic, and ultrasonic wedge. Each bonding method has its own advantages and disadvantages. Each method can be used in the process of the present invention. All result in a high quality performance inductor of the present invention. The manufacturing yield and tolerance can be easily controlled by off-the-shelf automatic wire bonding machines with loop control capability which are used for high density VLSI packaging. Circular bond wire loops with a minimum 65 μm separation can be repeatedly manufactured within few percentage geometrical variations using the polygonal movement of a wire bonding machine. Loop heights are also controllable.

The inductor of the present invention can be fabricated in Very Large Scale Integration (VLSI) and Ultra Large Scale Integration (ULSI) designs for all semiconductor material; e.g. in silicon-based RF integrated circuits and in GaAs MMIC's. The vertical placement of the bond wire loop separates the electromagnetic fields from the conductive substrate and this effectively reduces the substrate effects. Also, the bond wire has a wide cross-section and long periphery, resulting in negligible ohmic resistance. Small parasitic capacitance with the ground plane is also achieved. With all of these positive conditions, the bond wire is an excellent loop conductor for high quality factor and self-resonant frequency on chip inductors. It is obvious that this technique will not only satisfy the fundamental requirements of semiconductor manufacturing, but also reduce production cost over the complex processes discussed in reference to the prior art.

Conventional IC processing is completed, including the fabrication of normal bonding pads. Now, the inductor of the present invention is to be fabricated. For example, FIG. **1***a* illustrates in side view a meander-shaped inductor of the present invention. Top conductive lines **20** are illustrated along with bonding pads **26**. It will be understood that semiconductor device structures, not shown, may underlie and be connected to the top conductive lines. Input/output connections **30** have been made to two of the bonding pads **26**, as shown. FIG. **1***b* illustrates the same inductor in plan view.

The individual inductor loop **34** consists of, for example, a 100 μm wide horizontal strip (this dimension can be varied) with, for example, 58 μm by 58 μm bonding pad size and a 12 μm pad-to-pad spacing, as shown in FIG. **1***b*. These dimensions can be varied depending upon the design rule and are given for illustration purposes only.

The bond wire inductor of the present invention has a wider range of inductance variations due to the loop numbers. That is, the more loops used in the inductor, the higher the inductance. The magnetic flux linkage between the bond wire loops **34** is more efficient than that of spiral conductors where the horizontal geometry of the spiral inductors limits the flux linkage to the smaller internal loops. The bond wire inductor of the present invention also has smaller parasitic capacitance between the loops and the ground plane because the bond wires are separated from the silicon surface. Therefore, self-resonant frequencies of the bond wire inductors are higher than those of the spiral inductors. The associated frequency dependence is also improved in the inductor of the invention.

The material of the bond wires **34** can be either copper or gold which have very low series resistance and are therefore excellent options for inductors.

One minor limitation of this design is the use of the top metal conductor strips **20** that connect the bond wires to form the inductor. At high frequency, due to the skin effect (i.e., the current is confined to flow only at the surface of the conductor at high frequency), the ohmic strip resistance of the metal will be increased by the square root of the frequency. Fortunately, this limiting electrical performance can be improved by replacing the strip conductors with wedge bond wires. This will effectively increase the quality factor. The wedge bond wire with the rotary head can form low profile short interconnections. Using only bond wires to form the inductor of the invention results in an inductor having very low ohmic resistance and, consequently, a greatly improved quality factor. No change is required to the normal VLSI process in making the inductor of the present invention.

FIG. **3***a* illustrates in side view and FIG. **3***b* illustrates in plan view a spiral inductor made using the bonding technique of the present invention. Top conducting strip **20**, bonding pads **26**, input/output connections **30**, and bonding wire **34** are shown as in FIGS. **1***a* and **1***b*. The fabrication techniques are the same in this embodiment. The only difference is in the shape of the inductor.

Likewise, FIG. **5***a* illustrates in side view and FIG. **5***b* illustrates in plan view a solenoid inductor made using the bonding technique of the present invention. Top conducting strip **20**, bonding pads **26**, input/output connections **30**, and bonding wire **34** are shown as in FIGS. **1***a* and **1***b*. The fabrication techniques are the same in this embodiment. The only difference is in the shape of the inductor.

It will be understood by those skilled in the art that the present invention is not limited to those embodiments shown in the drawing figures. The figures show only three of the many possible inductor shapes that can be made using the process of the invention. Many other inductor shapes can be made (for example hexagon, octagon, circular, triangular, etc) without departing from the spirit and scope of the present invention.

The inductor of the present invention is manufactured using current existing bonding techniques which has been cleverly applied and integrated into the present and future VLSI technology.

The wire bond inductor of the present invention can be further designed into the on-chip transformers. FIGS. **2***a* and **2***b* show a meander-shaped transformer, in side view and plan view, respectively. The first input/output connections **30** of a first inductor are shown. The second input/output connections **32** of a second inductor are shown. The two inductors together form the transformer. FIGS. **4***a* and **4***b* show a spiral transformer and FIGS. **6***a* and **6***b* show a solenoid transformer. These miniaturized high performance transformers will have significant contributions to the current VLSI and ULSI integration process. The design of the transformer is based on integrating two inductors. While the inductor has two input/output connections, a transformer has four.

The on-chip inductor of the present invention fabricated using bonding techniques provides high quality factor performance and low cost for all semiconductor materials. The high quality inductor of the invention will play an important role especially in developing high performance silicon radio frequency integrated circuits (RF ICs) and microwave monolithic integrated circuits (MMICs). The bonding technique of the present invention can be used for any shape and size of inductors and transformers.

While the invention has been particularly shown and described with reference to the preferred embodiments thereof, it will be understood by those skilled in the art that various changes in form and details may be made without departing from the spirit and scope of the invention.

What is claimed is:

1. A method of fabricating a transformer in the fabrication of integrated circuits comprising:

providing bonding pads over a semiconductor substrate;

providing wedge bond wire underlying and connecting said bonding pads;

making first input/output connections with two of said bonding pads;

making second input/output connections with another two of said bonding pads; and

forming a plurality of bond wire loops, each loop connecting each two of said bonding pads wherein said bond wire loops connecting to said first input/output connections form a first inductor and wherein said bond wire loops connecting to said second input/output connections form a second inductor and wherein said first and second inductors together form said transformer.

2. The method according to claim **1** wherein said bond wire loops comprise copper or gold.

3. The method according to claim **1** further comprising:

forming semiconductor device structures in and on said semiconductor substrate wherein said semiconductor device structures include topmost said wedge bond wires; and

forming said bonding pads overlying and contacting said topmost said wedge bond wires.

4. The method according to claim **1** wherein said first and second inductors have a meander, spiral, or solenoid shape.

* * * * *

5

Ownership and Duration of Protection of Intellectual Property Rights Related to Integrated Circuits

Introduction

As discussed in Chapter 4, integrated circuits are capable of being protected by a number of intellectual property rights that are similarly available for other products or processes, except for, perhaps, plant variety rights and geographical indications. To recapitulate, the intellectual property rights which are applicable to integrated circuits include (1) patents, (2) copyright, (3) layout-designs[1] of integrated circuits, and (4) trademarks. There are essentially two alternative overviews of discussing IP ownership, i.e., the first based on a respective treatment of the types of IP rights and the second based on the integrated circuit's product development flow. We chose to write this chapter according to the first.

For ease of reference, a simple tabulation of the different types of intellectual property applicable to integrated circuits and their brief description is summarized as follows:[2]

Type of IP	Definition	Subject Matter
Patents	A patent is a monopoly right given by a country to the owner of the invention to enable him/her to prevent others from using, copying, or making the invention without the owner's consent in the country in which he/she has obtained patent protection.	New, inventive step, and industrially applicable inventions
Copyright	Copyright is a bundle of rights given to creators of works to make sure that only they can use and reproduce what they have created for their own purposes. This protects works like novels, computer programs, plays, sheet music, and paintings. These rights enable copyright owners to control the commercial exploitation of their work.	Original works of authorship
Layout-designs of integrated circuits	The 3-dimensional character of the elements and interconnections of an integrated circuit. An integrated circuit is an electronic circuit in which the elements of the circuit are integrated into some medium, and which, as a whole, functions as a unit.	
Trademarks	A trademark is a sign used by a person in the course of business or trade to distinguish his goods or services from those of other traders.	Signs or symbols to identify goods and services

We discuss the various issues of ownership of patents in integrated circuits on page 114. Copyright ownership issues relating to integrated circuits, particularly in respect to software or firmware, is also discussed later in this chapter. Layout-designs of topographies are discussed, considering the trademark ownership applicable to completed integrated circuit products.

As seen in the preceding chapters, the reusability of units of integrated circuits (e.g., logic core providing logical functions, cells, or gate netlists) and the relative ease of establishing layout-design rights for integrated circuits may be complemented with the more rigorously procured patent rights (e.g., on the fabrication process of certain gate structures or silicon substrate features); copyright (e.g., a developer software tool to be used with the integrated circuit unit or block, or where a ROM component of the integrated circuit is encoded with software); and trademark in the form of a catchy and convenient name (e.g., *Stratix*™ is a trade name referring to a family of FGPAs by Altera Corporation) for the bundled form of IP blocks, IP cores or logic cores, or the finished product (*iPod*™ by Apple Inc.) ready for retail. It is imperative, therefore, for the proprietorship of each of these intellectual property rights to be established beforehand, in order for any reusability and licensing thereof to be gainful to the owners of the IP rights, as well as for the licensees or users of these IP blocks, besides protecting the finished consumer product.

Following the product development process is an alternative perspective for considering how to treat ownership issues. Product development may be said to commence with the definition of product specifications. At this stage, patent and copyright ownership may arise, depending on what the specifications are, i.e., whether they are merely industrial standards or customer wish lists at one end (where it is doubtful if any IP rights could arise), to certain technical specifications that are more ingenious or conceptually new (where it is possible for some IP rights to arise). This goes on to each of the subsequent stages of product development, i.e., when the product specification is further expanded to become a microchip or system architecture, and having that architecture worked into circuit designs, netlists, etc. Once the circuit design is completed and ready for simulation, certain models or processing vectors may be found to be patentable or copyrightable. Only when all of these are prepared for the photolithographic process to proceed to the physical manufacture of the integrated circuit, would layout-design of topography rights become available alongside copyright and patent protection. When the product that comprises the integrated circuit is finished as a consumer product and is ready to be shipped out, trademarks become available in addition to the aforesaid IP rights.

Discussion of IP ownership issues may be further complicated by the territorial nature of IP rights, i.e., each IP right is a separate right in each country, and ownership has to be secured and considered within each country. This territorial division thus compounds the existing intellectual property rights available for ownership of integrated circuits, as discussed in Chapter 4. In this chapter, we cover the topic of ownership of IP rights of integrated circuits by considering each type of IP in turn and endeavor to cover the law primarily in the context of Singapore.

Ownership of Patent

A patent is the monopoly right granted to the proprietor of an invention, and such exclusive rights allow the proprietor to make, dispose of, offer to dispose of, use, import, or keep the invention, whether for disposal or otherwise for a duration of usually up to 20 years, in which the invention could be a product or process.

The patent is distinct from the underlying invention for which the patent was granted. In some cases, an inventor may assign his patent if he has one or, if he has not applied for a patent, he may enter into an agreement to assign his invention. Further, an invention of an employee shall be taken to belong to the employer and the employer may apply for a patent.

In most countries, ownership of a patent belongs to the inventor, and if there are two or more inventors, to the joint inventors.[3] How this ownership of a patent can be transferred from the inventor to a third party by the operation of law or agreement between the parties is discussed in greater detail in subsequent sections.

Co-Ownership of Patent

For joint proprietors, ownership of the patent can take one of two forms. One, unless there is any agreement to the contrary, the joint proprietors co-own the patent jointly in an equal undivided share,[4] in which case each proprietor owns one hundred percent of the patent in undivided portions. This manner of co-owning a patent, unless subject to any agreement to the contrary, entitles the respective proprietors to benefit from their rights associated with co-owning the patent. Such rights are set out under Section 46(2) to 46(5) of the Patents Act, and in summary, entitle any of the proprietors to do any act in respect of the patent for his own benefit without the consent or the need to account to the other co-proprietors.

However, one important point to note is that the act of the proprietor referred to under Section 46(2) of the Patents Act does not cover the situation whereby one of the proprietors wishes to grant a license under the patent, or assign or mortgage a share of the patent to a third party. In such a case, the proprietor who wishes to license, assign, or mortgage the patent requires the consent to do so from the other co-proprietors. Therefore, if a person obtains a license for a patent (also known as a licensee) from one of the proprietors of the patent, and the latter did not seek consent for such licensing from the other proprietors, the licensee who is granted the licensed patent may be sued by the other proprietors for patent infringement.

The joint proprietors may also enter into an agreement to make arrangements that differ from the above default position as expressed by the law.[5] In this case, the joint proprietors co-own the patent on the basis of a tenancy-in-common, and the proprietors can agree to the share that each proprietor has in the patent, respectively.

By way of illustration, three proprietors, A, B, and C, may co-own a patent relating to a system and method of avoiding crosstalk in the design of an integrated circuit on the basis of a tenancy-in-common, and they agree that the ownership of the patent is divided in the following manner: A owns 35 percent of the patent; B owns 35 percent of the patent, and C owns the remaining 30 percent of the patent.

Invention Belonging to the Employer

The patent of an invention does not necessarily belong to the inventor. It is also possible for an invention to be owned by another party other than the inventor. One possibility is the case of an employee who invents an invention and files a patent for that invention. In Singapore, by virtue of Section 49 of the Patents Act, an invention made by an employee that falls within the nature and scope of Section 49 belongs to the employer. Hence, although the invention is made by the inventor, i.e., the employee, the ownership of the invention belongs to the employer.

Next, we evaluate the various circumstances in which an invention made by an employee belongs to an employer through the operation of Section 49 of the Patents Act. Section 49(1) of the Patents Act reads as follows, whereby the invention belongs to the employer if:

(a) the invention was made in the course of the normal duties of the employee or in the course of duties falling outside his normal duties, but specifically assigned to him, and the circumstances in either case were such that an invention might reasonably be expected to result from the carrying out of his duties; or
(b) the invention was made in the course of the duties of the employee, and at the time of making the invention, because of the nature of his duties and the particular responsibilities arising from the nature of his duties, he had a special obligation to further the interests of the employer's undertaking.

By way of illustration, if A is the software engineer at a company and, in the course of his normal duties, he is required to design software to computerize the company's financial data, unless the parties agreed otherwise, the application software invented by A for the purposes of computerizing the financial data shall belong to the company, i.e., the employer he works for, by operation of Section 49(1) of the Patents Act.

Similarly, the ownership of the application software shall also belong to the company if B, the account executive of the company, whom the company specifically assigns the duty of assisting A in the computerization of the company's financial data, also plays a part in inventing the application software. Although the normal duties of an account executive do not involve the invention of application software, in this case, B is specifically assigned the duty of assisting A in the computerization of the financial data, and it can reasonably be expected that an invention related to the computerization of the financial data has resulted from the carrying out of his duties. Thus, the invention by B also belongs to the company, i.e., the employer.

The operation of Section 49(1) of the Patents Act that stipulates that the invention belongs to the employer and not the inventor or employee, is subject to the fulfillment of one out of two important criteria set out under Section 50(1) of the Patents Act. One, the employee was mainly employed in Singapore at the time he made the invention; or two, he was not mainly employed anywhere or his place of employment could not be determined, but his employer had a place of business in Singapore to which the employee was attached.

Invention Belonging to the Employee

One way to avoid the situation in which an invention made by an employee belongs to his employer through the operation of Section 49 of the Patents Act is for the employee to specifically address the issue of ownership by way of a contract before the invention is made. It is always possible for an employee who is highly regarded, or well sought after in the industry, to negotiate and bargain with his employer such that his contractual terms stipulate that the ownership of any invention belongs to him, even though he is under employment. The above also illustrates the principle that ownership of an invention can be altered by way of an agreement, so that an invention can be contracted to belong to either the employee or employer.

Invention Belonging to a Third Party

A third party who is not an inventor or does not fall within the ambit of the above scenario can still be granted a patent if he is the successor-in-title of the inventor.[6] He may take his title from the owner by way of an assignment from the owner who may be transferring his rights in the patent application if he has already applied for a patent or merely his rights in the invention if the owner has not applied for a patent at the relevant time. An assignment of a patent application must be in writing or in the form of a deed and must be formally executed by all the parties. There are no such requirements for an assignment of an invention since it is not governed by the Patents Act or any other written law.

Duration of Patent Protection

The duration of patent protection is valid for a period of 20 years from the date of filing the patent application provided that the proprietor of the patent renews the patent annually by paying a renewal fee.[7] It is possible for this 20-year term to be extended if the proprietor of the patent establishes that there was an unreasonable delay by the patent registrar in granting the patent.[8]

Though not applicable to integrated circuits, it is noteworthy that the 20-year term of a patent may be extended on the grounds that the subject of the patent includes any substance which is an active ingredient of any pharmaceutical product. Additionally, an extension may be granted if there was an unreasonable curtailment of the opportunity to exploit the patent that was caused by the process of obtaining marketing approval for the pharmaceutical product. In this case, the patent may be extended for a maximum period of 5 more years.[9]

Ownership of Copyright

It is important to understand that copyright protects the form of expression rather than the general ideas, concepts, or information in the copyright material.

Before discussing the various issues related to copyright ownership, it is first important to distinguish the difference between owning the copyright subsisting in a work as opposed to owning the physical medium on which the copyright is recorded.

Copyright and Physical Medium

By way of illustration, when a person buys a user guide for a computer application such as Microsoft PowerPoint, the person owns the physical medium from which the user guide is made, but he does not become the owner of the copyright of the user guide. This explains the reason why the person will still infringe the copyright of the user guide if he makes an unauthorized copy of the user guide for his friend, even if he has bought and is the owner of the user guide. In summary, the buyer of the user guide only owns the physical medium from which it is made, but not the copyright to it, which still remains with the author and not the buyer of the user guide.

Use of Symbol © to Indicate Ownership of Copyright

Copyright subsists automatically in copyright works when these original works are created or published and the owner has not copied it from any other persons. Hence, there is no requirement to formally register the copyright with any authority or governmental body in Singapore[10] in order to seek legal protection or to commence a legal action against an infringing party.

The owner of the copyright has the exclusive right to do any of the following acts in relation to the copyright work:[11]

- Reproduce in material form, i.e., making copies
- Publish the work
- Perform the work in public
- Broadcast the work to the public
- Make an adaptation of the work

Although there is no legal requirement to indicate the ownership of copyright in a work, it is common practice for the owner to protect the copyright of his work by way of using the copyright symbol © or a simple copyright notice such as *© 2010 J. Ross Publishing, Inc.* or *Copyright belongs to J. Ross Publishing, 2010*. The owner named in the copyright notice shall be presumed to be the owner of the copyright, although this presumption can be rebutted by adducing evidence which shows the contrary.[12]

Generally, for copyright to subsist in the copyright works, one important connecting factor is that the author of the work is either a Singapore citizen or resident in Singapore[13] if the work is first published in Singapore. If the work is first published in a country outside Singapore, the work must be published in Singapore within 30 days of its first publication to be considered as the first publication in Singapore.[14]

One more step before our discussion on the ownership of copyright. One needs to establish the subject matter in which the copyright subsists, which can be broadly classified into two types of copyright works, i.e., whether they are *works* as defined under Section 7(1) of the Copyright Act which include literary, dramatic, musical, and artistic works, also collectively referred to as authors' works; and *the subject matter other than works*, also collectively referred to as neighboring or entrepreneurial works which include sound recordings, cinematographic films, television and sound broadcasts, cable programs, and published editions of works.

Ownership of Copyright in Literary, Dramatic, Musical, or Artistic Works

Though dramatic, musical, and artistic works do not apply to copyright protection of integrated circuits, it is the purpose of this section to describe each of these works in slightly more detail so as to explain some of the fundamental principles and concepts of how copyright subsists in these works. Any subsequent discussion in this chapter emphasizes only the types of copyright protection relevant to integrated circuits.

1. **Dramatic works** Dramatic work includes (a) a choreographic show or other dumb show if described in writing in the form in which the show is to be presented; and (b) a scenario or script for a cinematograph film.[15] Therefore, dances and mimes which are recorded by words or symbols, that is, reduced to writing or a material form, are capable of being protected by copyright. It is also accepted that dramatic work includes work that must have sufficient unity to be capable of performance.

2. **Musical works** Although musical work is not defined under the Copyright Act in Singapore, it is considered to be an arrangement of musical notes even if the musical work is unpleasant or badly created. It is also important to clarify the confusion associated with the two different and distinct types of copyright that subsist in a song. Inevitably, a song consists of the music and the lyrics, in which case there are two types of copyright that subsist in it, i.e., literary copyright in the lyrics and musical copyright in the music.

3. **Artistic works** Artistic work is defined as (a) a painting, sculpture, drawing, engraving, or photograph, whether the work is of artistic quality or not; (b) a building or model of a building, whether the building or model is of artistic quality or not; or (c) a work of artistic craftsmanship to which neither (a) nor (b) applies.[16] It is interesting to note that the Copyright Act in Singapore specifically excludes a layout-design or an integrated circuit, as the sui generis protection for integrated circuits is provided by the Layout-Designs of Integrated Circuits Act (Cap. 159A).[17] Without this specific exclusion in the Act, integrated circuits could be protected by copyright as a form of artistic work since each mask may be made by a photographic process on a film of material.

Some examples of artistic works include specification drawings of ships, a flowchart, and a drawing of a hand holding a pencil. It is necessary that artistic works be original with some effort and skill expended in creating them. However, the quality of the artistic work is not relevant.

4. **Literary works** We have intentionally left the discussion of literary works to last. Unlike the other three forms of authors' works, literary works are the only relevant type of authors' works that could apply to the protection of integrated circuits. Literary work is defined under the Copyright Act to include (a) a compilation in any form; and (b) a computer program.[18] For compilations, a conversion table and street directories have been held

by the courts to be protectable. Most relevant to integrated circuit protection is the firmware or source code relevant to an integrated circuit. In this case, the computer program or the firmware that is burned into the ROM is protected under copyright as a literary work.

Under Section 30(2) of the Copyright Act, the general legal position is that the author of the original literary, dramatic, musical, and artistic works shall be the copyright owner. With the exception of photographic works, whereby the Copyright Act defines the author of a photograph to be the person who took the photograph,[19] the author of literary, dramatic, musical, and artistic works is generally considered to be the person who created the original work by expending skill, labor, and effort in such expression.

This ownership position can, however, be altered should the author acquire copyright ownership in the course of his employment, or if the author is commissioned to create the work, or the new owner may acquire such rights from the author by way of an assignment. We now examine each of these methods of acquiring ownership of copyright in greater detail.

Ownership of Copyright under Employment

In the event that an employee creates a copyright work in the course of his employment, the first copyright owner of this work shall be the employer and not the employee, unless this arrangement is excluded or modified by agreement between the parties.[20]

By way of illustration, if A is an employee of a company, S Limited, and during the course of his employment at S Limited he creates a computer program for a software application, the copyright of the source code, which is a literary work, shall be owned by the employer, S Limited, and not by the employee.

Once again, the publication of newspapers and magazines or periodicals is irrelevant and not applicable to our discussion of the ownership of copyright in integrated circuits. In this case, the employer, who is the publishing company, owns only the copyright relating to the publication and reproduction of the author's work in the newspapers, magazines or periodicals.[21] All the other copyright in the newspapers, magazines, or periodicals still remains with the employee.

Ownership of Copyright for Commissioned Works

In the event that an author creates the copyright work under the commission of another party, the first copyright owner of such work shall be the commissioning party.[22] However, take note that this rule may be excluded or modified by agreement between the parties.[23]

Furthermore, the operation of the above rule, whereby the commissioning party is the first copyright owner, only applies to limited types of artistic works such as where the author is commissioned to draw a portrait or to take a photograph or to make an engraving of a favorite Chinese calligraphy. In these cases, the ownership of such works belongs to the commissioning party. This rule of vesting ownership in the commissioning party does not apply to the other forms of authors' work such as literary, dramatic, and musical works in which case the copyright still continues to be vested in the author of such works.

By way of illustration, if S Limited commissions a third party, B, to write a computer program for a software application, the copyright of the source code, which is a literary work, is owned by B, the author of the source code of the computer program, and the rule enunciated above does not apply.

By Way of Assignment

In addition to the various ways of vesting ownership of copyright subsisting in authors' works or entrepreneur works discussed above, the owner could also transfer his copyright, whether in whole or in part, to any third party by way of an assignment.

Joint Authorship

The issue of joint authorship in the creation of authors' works needs to be discussed. Joint authorship is common for the creation of authors' work and involves several persons collaborating. The Copyright Act defines *work of joint authorship* as a work that has been produced by the collaboration of two or more authors and in which the contribution of each author is not separate from the contribution of the other author or the contributions of other authors.[24]

To qualify as a joint author, the person must contribute and show substantial contribution to the creation of the work, and not just mere criticism or a basic fact or idea. The joint authorship can take the form of tenancy-in-common or joint tenancy whereby the former refers to joint authors owning the copyright in separate and distinct shares of the same work, and such shares may be unequal. Unlike tenancy-in-common, in a joint tenancy all the owners hold an equal and undivided share in the work.

Duration of Copyright Protection for Literary, Dramatic, Musical, and Artistic Works

The duration of copyright protection for literary, dramatic, musical, and artistic works is 70 years from the end of the year in which the author died.[25] If, however, the work is published after the death of the author, it lasts for 70 years from the end of the year in which the work was first published.[26]

Ownership and Duration of Copyright Protection for Entrepreneurial Work or Neighboring Works

The other main category of copyright works is collectively known as entrepreneurial works or neighboring works, in which case they include sound recordings, cinematographic films, television broadcasts and sound broadcasts, cable programs, and published editions of works.

Similarly, these rights are usually hardly applicable to the intellectual property protection offered to integrated circuits but for the sake of completeness, the following section shall broadly discuss the basic concepts and principles of the copyright subsisting in these works.

1. **Sound recording** The Copyright Act defines *sound recording* as the aggregate of the sounds embodied in a record and a *record* is defined as a disc, tape, paper, or other device in which sounds are embodied.[27] As such, MP3s, compact discs, and cassette tapes are all sound recordings. It is therefore interesting to note that when a record is made on CD by a recording company, three types of copyright subsist in the record: (1) the literary work for the song lyrics, (2) the musical work for the music, and (3) the sound recording.

Under Section 97 of the Copyright Act, the legal position is that the first copyright owner of the copyright subsisting in a sound recording shall be the maker of the sound recording. The maker shall be the person who owns the record at the time when the record embodying the recording was first produced.

The duration of copyright protection for sound recordings is 70 years from the end of the year of release of the sound recording.[28]

2. **Cinematographic films** *Cinematograph film* is defined under the Copyright Act to mean the aggregate of visual images embodied in an article or thing so as to be capable by the use of that article or thing: (a) of being shown as a moving picture; or (b) of being embodied in another article or thing by the use of which it can be so shown, and includes the aggregate of the sounds embodied in a soundtrack associated with such visual images.[29] Therefore, a movie that is embodied in a DVD or VCD is considered

to be a cinematographic film. In this case, it refers to the production of the movie embodying the sounds associated with the visual images.

Under Section 98 of the Copyright Act, the general legal position is that the first copyright owner of the copyright subsisting in a cinematographic film shall be the maker of the film. The maker shall be the person who undertakes to make the necessary arrangement for the making of the film.

The duration of copyright protection for a cinematographic film is 70 years from the end of the year of release of the cinematographic film.[30]

3. **Broadcasts** *Broadcast* is defined as a broadcast by the emitting or receiving, other than over a path that is provided by a material substance, of electromagnetic energy.[31] In this case, it includes both a television and a radio broadcast. Therefore, the broadcaster of a movie that appears on the TV screen enjoys separate copyright protection in such a broadcast.

Under Section 99 of the Copyright Act, the legal position is that the first copyright owner of the copyright subsisting in a broadcast shall be the person who is the holder of a broadcasting license and who makes the broadcast.

The duration of copyright protection for a broadcast is 50 years from the end of the year of making the broadcast.[32]

4. **Cable programs** A cable program is a program which is included in a cable program service, i.e., a service consisting of sending by means of a telecommunication system of sounds or visual images or both for reception, otherwise than by wire telegraphy, at two or more places in Singapore for the purpose of being presented there to members of the public.[33] Therefore, the cable program, such as a television program transmitted to subscribers of cable TV, is protected by copyright for the preparation and broadcast of such television and/or sound program by cable or wire.

Under Section 100 of the Copyright Act, the legal position is that the first copyright owner of the copyright subsisting in a cable program shall be the person providing the cable program service in which the cable program is included.

The duration of copyright protection for a cable program is 50 years from the end of the year in which the cable program is first included in the cable program service.[34]

5. **Published editions of works** In addition to the copyright that subsists in literary, dramatic, musical, and artistic works, when these works are published there exists a separate copyright in the published editions of the

work. This form of copyright protection covers the typographical format such as presentation and layout of the work in the publication.

Under Section 101 of the Copyright Act, the legal position is that the first copyright owner of the copyright subsisting in a published edition of a work shall be the publisher of the edition of the work.

The duration of copyright protection for the published edition of a work is 25 years from the end of the year in which the edition was first published.[35]

Ownership of Layout-Designs of Integrated Circuits

Before considering the ownership issues relating to the layout-design of an integrated circuit, it is necessary to trace the manner of formation of the layout-design to its creation stage, that is, when a person creates that layout-design. Generally, under the Layout-Designs of Integrated Circuits Act in Singapore, ownership of the layout-design is vested in the creator of the layout-design.[36] If he uses or is assisted by an electronic design automation (EDA) software application or computer-aided tool, or otherwise makes the arrangements to create the layout design, use of such tools or assistance does not generally diminish his role as the creator of the layout-design. In fact, according to the definition contained in Section 2 of the Layout-Designs of Integrated Circuits Act, the creator will still be the person who made such arrangements for the creation of the layout-design. There are two exceptions to the above general rule of ownership of the layout-design: the creation of the layout-design by way of commission or during the course of employment with an employer.[37]

To seek legal protection for the layout-design, there is no system of registration required in Singapore and the rights granted to the owner by the Layout-Designs of Integrated Circuits Act include his right to make copies and to commercially exploit the protected layout-design.[38]

Qualifications for Creatorship and Ownership

However, it is important to note that the creation of a layout-design of an integrated circuit by a creator does not automatically qualify the work for protection and, hence, does not automatically create a layout-design right of ownership. Depending upon jurisdiction, generally speaking the following criteria should be met in the creation process for the layout-design to qualify for creatorship and, in turn, ownership:[39]

1. **Originality** Originality arises where the layout-design is the result of the creator's own intellectual effort. In other words, suppose A happens to know how to operate the EDA application or tool but is faithfully taking detailed instructions written by C in creating the layout-design. In this case, A cannot be considered as the creator. This is analogous to a person typing out a handwritten manuscript of an author who does not know how to operate a typewriter. Hence, by tracing the originality and intellectual effort to C, she would qualify as the creator.

2. **Not commonplace** The layout-design must not be commonplace among your peers or fellow artisans, that is, other creators of layout-designs or integrated circuit manufacturers. This criterion may be further qualified as follows:

- Whether or not a layout-design is commonplace is to be considered at the time of its creation. Hence, one cannot use retrospection or hindsight to deny the novelty of a layout-design.
- If the layout-design is comprised of elements or interconnections that are commonplace in the semiconductor industry, the combination of the elements or interconnections may be taken as a whole in considering whether it meets the requirements of originality and non-commonplace.[40]

3. **Independent creation** Similar, and even identical, layout-designs may all enjoy protection if they have been created independently, regardless of when they are created. Some countries prescribe a starting date upon which such a criterion applies.[41] In other words, independent creation occurs if one has (a) expended sufficient intellectual effort to meet the originality requirement, and (b) arrives at a layout-design that is not commonplace in the semiconductor industry but is exactly the same as an earlier layout-design made by someone else (that is, the existence of just one other identical layout-design is not considered to have established that the particular layout-design is commonplace), and (c) has independently created the layout-design, i.e., having achieved both (a) and (b) independently.[42]

4. **Reduction to a material form** The layout-design is only considered as having been created when it has been reduced to or fixed in a material form,[43] which may include incorporating that layout-design into an integrated circuit. One cannot say that he has his layout-design in his mind and claim protection therefore on that basis.

5. **Exclusion of non-topographical aspects of the layout-design from protection** Exclusion may result if the layout-design also discloses additional subject matter such as illustrating or embodying a process, method, concept or principle; for example, if the layout-design illustrates a method of monitoring or regulating power supply, that method cannot be considered protected under the layout-design law and must be considered under other laws, e.g., patent law, if protection over that method is desired. In other words, the precise nature of protection for an integrated circuit under the layout-design laws is only for its topographical features.

The above five qualifications concern the creation of the layout-design itself. There is one further qualification to be met concerning the status of the creator or owner. Bearing in mind that the ownership of the layout-design may remain with the creator or pass on to a company which employs him to create that layout-design, depending on the situation, we shall explain the status requirement as follows:

6. **Nationality or residency requirement** The owner must be a national of, domiciled or resident in Singapore, or have a commercial establishment in any of the countries which are party to the TRIPS Agreement. The owner may be a natural person (i.e., an individual person) or a legal entity (e.g., a corporation, statutory body, etc., or the government) of any of the WTO member countries.[44]

When the creation of a layout-design fulfills qualifications (1) to (6) above, then the layout-design is said to be qualified for protection under the layout-design law. It would then be worthwhile to consider its ownership.

Commissioning

The owner of a layout-design as mentioned earlier shall generally be the creator of the layout-design, except for cases when the layout-design is created in pursuance of a commission or is created in the course of employment.[45]

In the case in which the layout-design is created pursuant to a commission by a person, then the person who commissioned the layout-design to be created will be the owner of the layout-design.[46] By way of illustration, if Design House Limited commissions B to create a layout-design for a new integrated circuit to be used as a microprocessor that they intend to sell, even though B creates the layout-design, the owner continues to be Design House Limited, the party who commissioned the creation of the layout-design.

Employment

In the case in which the layout-design is created in the course of employment, then the employer will be the owner of the layout-design created.[47]

If C is an employee of Design House Limited and C creates a layout-design for a new integrated circuit, the owner of the layout-design, therefore, shall be Design House Limited and not C, the employee.

Duration

The layout-design of an integrated circuit created after the enactment date of the Layout-Designs of Integrated Circuits Act, February 15, 1999, will be protected for a period of 10 years after the layout design is first commercially exploited, provided that it is commercially exploited within five years of its creation.[48] In any other case, the layout design of the integrated circuit is protected for a period of 15 years after its creation.[49]

Ownership of Trademark

Introduction

The proprietor of a trademark can be a person, a sole proprietorship, a partnership, or a company. Unless the trademark is being assigned, the proprietor is generally also the applicant for a registered trademark. Although it is not mandatory in Singapore for a trademark to be registered, registering a trademark gives the proprietor rights set out under the Trade Marks Act, including the exclusive rights to use the trademark and authorize other persons to use the trademark. Furthermore, the proprietor can enforce his legal rights by way of injunction and/or by claiming for damages and relief from the infringer for his acts of trademark infringement.

Use of Symbol ® to Represent a Registered Trademark

Once a trademark is registered, the proprietor may use the symbol ® next to the trademark to indicate that it is registered so that it serves to notify others that the proprietor has claims over the trademark. For example, Intel® and AMD® are registered trademarks owned by Intel Corporation Inc. and Advanced Micro Devices Inc., respectively, in Singapore. Therefore, if a person falsely represents a mark as being a registered trademark such as by using the symbol ® next to the mark when the mark is in fact not registered, pursuant to Section 51 of the Trade Marks Act, he could be guilty of an offense and, on conviction, the offender shall be liable to a fine not exceeding S$10,000.

Unregistered Trademark

We discussed the use of the symbol ® to represent a registered trademark. However, if a trademark is unregistered for whatever reasons, the proprietor can use the symbol ™ to indicate his ownership of the mark. There are many reasons for the mark being refused registration and some of the common grounds for refusal of registration are as follows:

- Mark which does not qualify as a sign for the purposes of the definition of trademark
- Mark which is devoid of any distinctive character
- Mark consists exclusively of signs which describe the kind, quality, intended purpose, value, geographical origin, or other characteristics of goods
- Mark which is contrary to public policy or morality
- Application of the mark is made in bad faith
- Mark conflicts with earlier trademarks and earlier rights

Co-ownership of Trademark

In accordance with Section 37 of the Trade Marks Act, unless there is any agreement to the contrary, if there are two or more proprietors of a trademark, then the co-proprietors co-own the trademark jointly in an equal undivided share, in which case each proprietor owns one hundred percent of the trademark in undivided shares.

The above manner of co-owning a trademark, unless subject to any agreement to the contrary, entitles each proprietor to benefit from his rights associated with co-owning the trademark. Such rights are set out under Sections 37(3) to 37(5) of the Trade Marks Act, which in summary entitles any of the proprietors of the trademark to do any act in respect of the registered trademark for his own benefit without the consent or the need to account to the other co-proprietors.

However, one important point to note is that the act of the proprietor referred to in Section 37 of the Trade Marks Act above does not cover situations in which one of the proprietors wishes to grant a license, assign or charge a share of the registered trademark to a third party. In such a case, the proprietor who wishes to license, assign, or charge the registered trademark requires the consent to do so from the other co-proprietors. As such, if a person obtains a license for a registered trademark (also known as a licensee) from one of the proprietors of the trademark and the latter did not seek consent for such licensing from the other proprietors, the licensee who is granted with the licensed registered trademark may be sued by the other proprietors for trademark infringement.

As indicated above, the joint proprietors of the registered trademark may enter into an agreement to make arrangements that are different from the above default position as expressed by the law. In which case, the joint proprietors co-own the registered trademark on the basis of a tenancy-in-common and the proprietors can agree to the share each proprietor has in the registered trademark, respectively.

By way of illustration, three proprietors A, B and C, may co-own a registered trademark on the basis of a tenancy-in-common, and they agree that the ownership of the registered trademark shall be divided in the following manner: A owns 10 percent of the registered trademark; B owns 20 percent of the registered trademark and C owns the remaining 70 percent of the registered trademark.

Ownership of Trademark by Way of Assignment

Section 36 of the Trade Marks Act provides that a registered trademark is personal property and the trademark is, therefore, capable of being assigned and transmitted in the same way as other personal or movable property. It is also important to note that the assignment or transmission of a registered trademark may be partial, which means that the assignment or transmission only applies to some but not all the goods for which the trademark is registered. For an assignment or an assent of the registered trademark to be effective, it must be in writing and signed by the party assigning the trademark, also known as the assignor, or his personal representative. If the assignor or personal representative is a corporate body, the assignment or assent must be affixed with the company seal.

Although the registration of an assignment of a trademark is not mandatory under the Trade Marks Act, and its nonregistration does not affect the validity of the trademark nor the assignment, Section 39(3) of the Trade Marks Act provides that the assignment will be ineffective as against a person acquiring a conflicting interest in the registered trademark in ignorance of the previous transaction.

Assignment of Unregistered Trademarks

The Trade Marks Act does not affect the assignment or transmission of an unregistered trademark as part of the goodwill of a business. As such, this would mean that the assignment and transmission of an unregistered trademark is governed by common law. Under common law, the right of property did not reside in the mark but in the goodwill of the business in which the mark is used. Hence, under common law, there is no separate and independent right of property in the trademark; it is an inseparable part of the goodwill of the business.

Duration of Trademark Protection

The duration of trademark protection for its proprietor is for a period of 10 years from the date of registering the trademark with the Trade Mark Registry.[50] The trademark can thereafter be renewed on a 10-year basis by payment of renewal fees. Hence, so long as renewal of the trademark is made every 10 years, the trademark can exist in perpetuity. One example of an old trademark registered in Singapore is Coca-Cola, which has been registered since May 13, 1939, i.e., the Coca-Cola trademark has been around for the past 70 years.

Summary

Integrated circuits are a form of intellectual property capable of being protected by a number of intellectual property rights related to ownership rights under patent, copyright, layout-designs of integrated circuits, and trademark law.

Patent ownership grants the proprietor of an invention the monopoly right to make, dispose of, offer to dispose of, use, import, or keep the invention, whether for disposal or otherwise, for a duration usually of up to 20 years. This term may be extended if the proprietor establishes that there was an unreasonable delay by the patent registrar in granting the patent. A patent can be co-owned by joint proprietors, pursuant to Sections 46(2) to 46(5) of the Patents Act. It is also possible for a patent of an invention to be owned by another party other than the inventor. One such example is by virtue of Section 49 of the Patents Act that provides that an invention made by an employee that falls within the nature and scope of Section 49 of the Patents Act, belongs to the employer. In addition, ownership of a patent can also be transferred to another person other than the inventor by way of an assignment.

Copyright subsists automatically in copyright works when original works are created or published and the owner has not copied it from any other persons. There is no requirement to formally register the copyright with any authority in Singapore to seek legal protection or to commence a legal action against an infringing party. It is common practice for a copyright owner to protect the copyright of his work by using the copyright symbol © or a copyright notice, although this can be rebutted by adducing evidence to the contrary. The owner of the copyright has the exclusive rights to (1) reproduce in a material form, i.e., make copies; (2) publish the work; (3) perform the work in public; (4) broadcast the work to the public; or (5) make an adaptation of the work. The duration of copyright protection for literary, dramatic, musical, and artistic works is 70 years from the end of the year in which the author died. If, however, the work is published after the death of the author, it lasts for 70 years from the end of the year in which the work was first published.

If an employee creates a copyright work in the course of his employment, the employer shall be the first copyright owner of the work and not the employee, unless the parties agree otherwise. If an author creates the copyright work under the commission of a commissioning party, the commissioning party shall be the first copyright owner of such work, unless otherwise agreed by the parties. This rule of vesting ownership upon the commissioning party applies only to certain forms of artistic works and does not apply to literary, dramatic, and musical works in which case the copyright still continues to be vested with the author of such works. Additionally, a copyright owner may transfer his rights by assigning his ownership to another party.

Ownership of a trademark registered under the Trade Marks Act gives the proprietor exclusive rights to use the trademark, authorize other persons to use the trademark, and to enforce legal rights by way of injunction and/or by claiming for damages and relief from the infringer for his acts of trademark infringement. The duration of trademark protection for its proprietor is 10 years from the date of registering the trademark with the Trade Mark Registry. The trademark can thereafter be renewed on a 10-year basis by payment of renewal fees.

The proprietor of a trademark is generally also the applicant for a registered trademark unless the trademark is assigned. Under Section 36 of the Trade Marks Act, for an assignment or an assent of the registered trademark to be effective, it must be in writing and signed by the party assigning the trademark, also known as the assignor, or his personal representative. If the assignor or personal representative is a corporate body, the assignment or assent must be affixed with the company seal. Although registration of the assignment is nonmandatory under the Trade Marks Act, Section 39(3) of the Trade Marks Act provides that the assignment will be ineffective against a person acquiring a conflicting interest in the registered trademark in ignorance of the previous transaction.

Once a trademark is registered, the proprietor may use the symbol ® next to the trademark to indicate that it is registered. If a trademark is unregistered for whatever reason, the proprietor can use the symbol ™ to indicate his ownership of the mark.

For layout-designs of integrated circuits, generally under Section 2 of the Layout-Designs of Integrated Circuits Act in Singapore, ownership of the layout-design is vested in the creator of the layout-design. There are two exceptions to this general rule of ownership: (1) creation of the layout-design by way of commission, or (2) creation in the course of employment. The rights granted to the owner of a layout-design include rights to make copies and to commercially exploit the protected layout-design. There is no system of registration in Singapore. A layout-design will be protected for 10 years if it is first used commercially within five years of creation, or in any other case, 15 years after its creation.

Creation of a layout-design by a creator does not automatically qualify the work for protection and, hence, does not automatically create a layout-design right of ownership. Depending on jurisdiction, the following criteria should be met: (1) the layout-design must be original and a result of the creator's own intellectual effort; (2) the layout-design is not commonplace; (3) the layout-design must be developed independently; (4) the layout-design must be reduced or fixed in a material form, such as incorporation into an integrated circuit; (5) the layout-design must exclude all nontopographical aspects of the layout-design from protection; and (6) the owner must be a national of, domiciled or resident in Singapore, or have a commercial establishment in any of the TRIPS member countries.

6

Infringement of Intellectual Property Rights Applicable to Integrated Circuits

Introduction

Chapter 5 discusses the ownership and duration of protection of the various intellectual property rights that are applicable to integrated circuits' protection in detail. It is equally important for a proprietor of such integrated circuits to understand, after seeking legal protection for those applicable intellectual property rights, in the event of an intellectual property infringement by a third party, what the legal actions are that he/she can take as a proprietor of the integrated circuits, and what legal remedies are also available when there is such an intellectual property infringement.

In this chapter, therefore, the discussion is focused primarily on the infringement of such intellectual property rights by a third party. Similar to the previous chapters, the discussion in this chapter is circumscribed to only those intellectual property rights that can apply to integrated circuits. Regarding copyright infringement, the emphasis is restricted to copyright infringement of literary works, that is, the computer program that is relevant and applicable to integrated circuits.

To facilitate reading of this chapter, it is worthy to note that the chapter is subcategorized into (i) patent infringement; (ii) copyright infringement; (iii) layout-design of integrated circuits' infringement; and (iv) trademark infringement followed by a brief summary of all the essential legal principles enunciated in the above subcategories. The second section of this chapter discusses patent infringement.

Patent Infringement

Once a patent is granted, the proprietor of the patent is entitled to exploit the patent to the exclusion of others. With this exclusive monopoly also comes the right to sue any offending party for patent infringement.

Definition of Patent Infringement

Under Singapore law, what amounts to an infringement of a patent is defined in Section 66 of the Patents Act (Cap 221). To sue for patent infringement, the patent proprietor must prove that an act of infringement, as defined in Section 66 of the Patents Act, has been committed by the infringer. It is noteworthy that there is a twofold test involved in the process of determining whether there is a patent infringement.

First, the patent proprietor (or the plaintiff of the suit, in this case) must show that the infringer (or the defendant of the suit, in this case) commits one of the prohibited *acts* as set out in Section 66 of the Patents Act:

A person infringes a patent for an invention if, but only if, while the patent is in force, he does any of the following things in Singapore in relation to the invention without the consent of the proprietor of the patent:

(a) where the invention is a product, he makes, disposes of, offers to dispose of, uses or imports the product, or keeps it whether for disposal or otherwise;
(b) where the invention is a process, he uses the process or he offers it for use in Singapore when he knows, or it is obvious to a reasonable person in the circumstances, that its use without the consent of the proprietor would be an infringement of the patent; or
(c) where the invention is a process, he disposes of, offers to dispose of, uses or imports any product obtained directly by means of that process or keeps any such product whether for disposal or otherwise.

To illustrate the acts of patent infringement, consider that Company X owns a patent for in vitro detection of antibodies. Test kits incorporating the same process of detection that are manufactured and sold by Company Y without Company X's consent would amount to a patent infringement by Company Y.[1] Similarly, where Company A owns a patent for a data storage device, Company B selling and marketing the same device under another name would amount to patent infringement by Company B.[2]

Second, the patent proprietor must prove that the infringing product or process falls within the claims of the patent.[3] For example, in the Singapore case of Institut Pasteur & Anor v. Genelabs Diagnostics Pte Ltd & Anor (2000) SGHC 53, the plaintiffs of this case had to prove that the defendants' diagnostic test kits infringed that particular claim in their specification of the patent relating to the processes for the in vitro detection of, among other things, HIV-2 antibodies.

Defenses to Patent Infringement

It is also important to highlight what does not amount to patent infringement under the Patents Act. These exclusions to patent infringement are set out under Sections 66(2)(a) to (i) of the Patents Act. That is, an act which would normally constitute patent infringement of an invention will not be so if the invention is used for any of the following:

(a) private noncommercial acts;

It is noteworthy that if the act has a dual purpose and one of those purposes is commercial in nature, then the exception will not apply. See the English decision *Smith, Kline & French Laboratories Ltd v Evans Medical Ltd (1989) F.S.R. 513.*

(b) experiments;

The English Court of Appeal has held that the word *experiment* is to be given its ordinary meaning. See *Monsanto Co v Stauffer Chemical Co (1985) R.P.C. 515, per Dillon L.J.*

The Japanese Supreme Court has interpreted a similar provision in Japanese patent law to allow clinical testing of drugs during the term of the patent, such that generic drugs can be available on the market as soon as the patent expires. See *Ono Pharmaceuticals Co Ltd v Kyoto Pharmaceutical Industries, Ltd (Judgment dated 16 April 1999).*[4]

NB. This is directly related to Section 66(2)(h) of the Patents Act, whereby this section permits the use of a patented invention for clinical testing to meet requirements for marketing approval within Singapore.[5]

(c) extemporaneous dealings with a medicine for an individual;

Briefly, the extemporaneous preparation of a medicine for an individual in a pharmacy and in accordance with a prescription that has been provided by a registered medical or dental practitioner does not amount to infringement. Dealings with medications prepared in this manner are similarly deemed to be noninfringing.[6]

(d) use of inventions forming a part of an air or sea vehicle that is temporarily within the territory of Singapore;

Section 66(6) of the Patents Act defines *relevant aircraft, hovercraft,* or *vehicle* as an aircraft, a hovercraft or a vehicle registered in, or belonging to, any country, other than Singapore, which is:

(i) a party to the Paris Convention; or
(ii) a member of the World Trade Organization.

Looking at English case law for guidance, *temporarily* has been interpreted as meaning transient or for a limited period of time, not depending on frequency. See *Stena Rederi AB v Irish ferries Ltd (2002) R.P.C. 50* (at p. 990) and *(2003) R.P.C. 36* (at p. 668). *Accidental* entry is thought to occur when a ship is lost or blown off course in bad weather.[7]

(e) use of inventions exclusively for the needs of the relevant ship, where the ship has entered the territory of Singapore;

Section 66(6) of the Patents Act defines *relevant ship* as a ship registered in, or belonging to, any country, other than Singapore, which is:

(i) a party to the Paris Convention; or
(ii) a member of the World Trade Organization.

It is emphasized that use of the word *exclusively* distinguishes use of the invention specifically for the needs of the ship, versus use for a job that the ship has been employed to do. By way of illustration, as adapted from *Terrell on the Law of Patents*,[8] where a ship is used for laying marine cables in Singapore, this operation would not be protected by Section 66(2)(e) of the Patents Act because it is not being performed for the exclusive needs of the ship, but rather for the party who has employed the ship to lay the cables. In contrast, it may be contemplated that if the invention was being used to prevent a ship from sinking in Singapore, then perhaps this would constitute use for the *exclusive needs* of the ship and would fall within the exception afforded by Section 66(2)(e) of the Patents Act.

(f) use of inventions that are part of an *exempted aircraft*, for example an aircraft in Singapore for the purpose of a commercial flight;

Section 66(6) of the Patents Act defines *exempted aircraft* as an aircraft to which Section 5 of the Air Navigation Act (Cap. 6) applies. Following the interpretation of similar provisions in English statute law, this exception provides protection in respect of inventions that are part of exempted aircrafts which are lawfully entering or crossing Singapore, and of importation into Singapore, or the use or storage, of any part or accessory of such aircrafts.[9]

(g) dealings with an invention for which a valid license has been given for its production;

This is the general provision on parallel importation and applies to any patented invention other than a pharmaceutical product. Broadly, it is not an infringement of a patent to import Singapore a patented product into, or a product obtained from a patented process, if the product was produced by the patent proprietor or another party who is licensed by the proprietor.[10] There are three more qualifications to this provision:[11]

(i) importation is permitted even if the patent proprietor in Singapore and the patent proprietor in the country of manufacture are different;
(ii) the provision applies to the importation of a patented product, or a product produced by a patented process, but *not* to the use of a *patented process* in Singapore (See: *Dextra Asia Co Ltd v Mariwu Industrial Co (S) Pte Ltd [2006] 2 SLR 154.*); and
(iii) for the purposes of determining whether the patented product was produced with the proprietor's consent, any conditions imposed by the patentee restricting the resale of the product outside the territory of manufacture will be disregarded. This is known as *deemed consent.*

The scope of this section is not to be interpreted so broadly as to imply that importation of pharmaceutical products produced in other countries under compulsory license is allowed.[12]

(h) acts to support an application for marketing approval for a pharmaceutical product; or

This section permits the use of a patented invention for clinical testing to meet requirements for marketing approval for a pharmaceutical product within Singapore.[13] Please see also subparagraph (b) experimental purposes, above.

(i) dealings with a pharmaceutical product for a specific patent in Singapore.

Section 66(2)(i) of the Patents Act provides an important exception to patent infringement, whereby importation of a drug into Singapore is permitted prior to the product being put out on the local market, where it is required for use by a particular patient and the Health Science Authority has granted the relevant approval. This approval is granted specifically for the importation of the pharmaceutical drug for a particular patient. In this way, infringement is circumvented through authorized parallel importation.[14]

Though note Section 66(3) of the Patents Act operates only while the pharmaceutical product is not available in Singapore, and once a drug company who is the patent proprietor sells the drug in Singapore, a parallel importer can bring in the drug from abroad.[15]

In summary, these acts as described above would therefore provide defenses to a claim of patent infringement.

Remedies for Patent Infringement

When a patent has been infringed, the patent proprietor may seek legal recourse by commencing legal proceedings against the patent infringer.[16] In this regard, the patent proprietor may seek one of the following remedies as prescribed under Section 67 of the Patents Act:

(a) for an injunction restraining the defendant from any apprehended act of infringement;
(b) for an order for him to deliver up or destroy any patented product in relation to which the patent is infringed or any article in which that product is inextricably comprised or any material and implement the predominant use of which has been in the creation of the infringing product;
(c) for damages in respect of the infringement;
(d) for an account of the profits derived by him from the infringement; and
(e) for a declaration that the patent is valid and has been infringed by him.

It should also be noted that besides the patent proprietor, the holder of an exclusive license can also sue in his own name for patent infringement committed after the commencement of the license.[17] To avoid any evidential difficulties in the pursuit of a legal remedy, the exclusive licensee should register his status as the exclusive licensee of the patent as set out in Section 75 of the Patents Act.[18]

Damages and Account of Profits

Pursuant to Section 67(1) of the Patents Act, a plaintiff can elect either 6.2.3(c) damages or 6.2.3(d) an account of profits, as set out above, but not both as the Court shall not, in respect of the same infringement, award damages to the proprietor of a patent and also order that he shall be given an account of the profits.[19] This is to prevent overcompensation to the plaintiff.

The election is usually carried out before an inquiry to assess damages or account of profits, and this is well illustrated in the Singapore case of *Main-Line Corporate Holdings Ltd* v. *United Overseas Bank Ltd and another (2007) 1 SLR 1021*. The plaintiff, Main-Line Corporate Holdings, was the proprietor of a patent in respect of a method and system to determine the operating currency in which to process a transaction for a credit card, a charge card, or a debit card at the point of sale between a merchant and the holder of the relevant card. The plaintiff alleged that the defendants' (United Overseas Bank and First Currency Choice Pte) technology known as the First Currency Choice System (*FCC System*) performed the same functions as the plaintiff's patent and therefore infringed its patent.

The High Court, in this case, granted judgment in favor of the plaintiff on the grounds that the patent was valid and had been infringed. The Court also granted an injunction against further infringement by the defendants. The Court further acknowledged that there would be an inquiry by the patent registrar on damages or an account of profits pursuant to Section 67(2) of the Patents Act, and that the plaintiff was to make its election at or before the stage of directions to be given for such an inquiry.

For recovery of damages under Section 69 of the Patents Act, there are also certain restrictions of which the plaintiff should be aware. For instance, damages will not be awarded and no order for an account of profits will be made in cases:

(a) against a defendant who proves that at the date of the infringement he was not aware, and had no reasonable grounds for supposing, that the patent existed;[20]
(b) if renewal fees or annuities for the patent have not been paid;[21]
(c) where the patent was granted on the basis of any Examination or Search and Examination Report provided by the patent proprietor during the patent application process under Section 29 of the Patents Act and the claim which is alleged to have been infringed was not a claim that was examined or referred to in the above Report;[22]
(d) where the patent was obtained on the basis of a corresponding application and the claim which is alleged to have been infringed has not been examined to determine whether the claim appears to satisfy the criteria of novelty, inventive step and industrial applicability;[23] or
(e) where an amendment of the specification of the patent has been allowed, in respect of infringement committed before the decision to allow the amendments, unless the Court is satisfied that the specification of the patent as published was framed in good faith and with reasonable skill and knowledge.[24]

The operation of Section 69(1) of the Patents Act is well illustrated in the Singapore case of *Seiko Epson Corp* v. *Sepoms Technology Pte Ltd and another (2007) 3 SLR 225*. In this case, the plaintiff commenced a suit on October 5, 2005, against the defendants for infringement of one of their patents. On August 2, 2006, the parties reached a consent judgment in which the defendants admitted liability for the infringement of the patent and agreed to an account of profits. The defendants acted accordingly and filed in their accounts commencing from the date of the writ (October 5, 2005), as they had only first acquired knowledge of the patent on October 7, 2005, when they were served with the statement of claim in the suit. Prior to the suit, the defendants had not received any *cease and desist* letter from the plaintiff, nor were the defendants informed orally or in writing of the existence of the patent by the plaintiff or of the defendants' infringement. Thus, the defendants were of the view that they were only liable to account to the plaintiff from October 7, 2005, to August 2, 2006. The High Court agreed with the defendants and held that Section 69(1) of the Patents Act limited the defendants' liability for damages for infringement and/or account of profits to the period when they knew or had reasonable grounds for supposing that the infringed patent existed. It was, therefore, well within the rights of the defendants to assert and subsequently prove at the inquiry that they had no knowledge of the existence of the patent until they were served with the writ in the suit.

Costs

Apart from claiming for damages or an account of profits as a way of seeking remedy for patent infringement, a plaintiff can also claim for the costs of the action he commenced, pursuant to Order 59 of the Rules of Court.

Procedural Issues

An action for patent infringement must be commenced in the High Court as prescribed by the Patents Act.[25] A special IP Court has been established in the High Court to hear intellectual property disputes.[26] Legal procedures pertaining to patent infringement actions are governed by Order 87A of the Rules of Court.

Copyright Infringement

Copyright ownership entitles the copyright owner to enjoy certain exclusive rights in literary works (which include computer programs).[27] These rights are set out in Section 26(1)(a) of the Copyright Act (Cap 63) and they include:

(a) to reproduce the work in a material form;
(b) to publish the work if the work is unpublished;
(c) to perform the work in public;
(d) to communicate the work to the public;
(e) to make an adaptation of the work; and
(f) to do any of the above acts in relation to an adaptation of the work in issue.

Definition of Copyright Infringement

Under the Copyright Act, there are three ways in which liability for copyright infringement can arise for a copyright infringer:

(a) by doing in Singapore without the license of the copyright owner, one or more of the exclusive acts reserved to the copyright owner (also known as primary or direct copyright infringement);[28]
(b) by authorizing primary copyright infringement; or
(c) by the commercial exploitation of known infringing copies (also known as secondary or indirect copyright infringement).[29]

Following the above, this section discusses the details of how copyright infringement can occur under the following three main headings:

(I) Primary Infringement;
(II) Authorizing Infringement; and
(III) Secondary Infringement.

(I) Primary Infringement

For primary infringement, copyright is considered infringed when a person does in Singapore any act comprised in the copyright without the license of the copyright owner.[30] Section 9(1) of the Copyright Act describes *an act comprised in the copyright* as any of the acts that the copyright owner has the exclusive right to do which are set out in Section 26(1)(a) of the Copyright Act as discussed above.

For the Court to determine whether the work has been infringed under Section 31(1) of the Copyright Act, the following requirements must be met:

(a) the infringing act must have been committed in Singapore, also known as the *territoriality* principle;

(b) the infringer must have copied the copyright owner's work, also known as the *causal connection* principle; and
(c) the infringing act relates to a substantial part of the copyright work.

Under Sections 17(a) and 17(b) of the Copyright Act, any references to the reproduction of any work in a material form, shall include references to the storage of that work in a computer or on any medium by electronic means. Since Section 7A of the Copyright Act includes computer programs within the meaning of literary works, this means that the storage of any computer programs in a computer, whether the storage is on a hard disk drive or a Read Only Memory chip (ROM chip), could result in copyright infringement.

The Singapore case of *Aztech Systems Pte Ltd* v. *Creative Technology Ltd (1996) 1 SLR 683* is an example that provides a good illustration of how legal principles for copyright infringement are being applied by the Courts. In this case, the High Court considered the issue of copyright infringement through storage of computer programs in a computer. This case concerned copyright in computer programs connected with Sound Blaster sound cards developed by Creative Technology (*Creative*). Aztech Systems Pte Ltd (*Aztech*) developed Sound Galaxy sound cards which were compatible or interoperable with the application programs that had been developed to operate with Sound Blaster. Creative alleged that in developing the Sound Galaxy sound cards, Aztech infringed its copyright of the software programs in the firmware of Creative's chip by copying the whole or parts of Creative's software programs.

Two types of copyright infringement were alleged by Creative: (a) copying by disassembly; and (b) copying by simply running the software program. The High Court dismissed Creative's claims and ruled that although there was no doubt that Creative's program TEST-SBC had been copied by Aztech, there was insufficient evidence that any other software program was similarly copied by disassembly in the course of Aztech's investigations into Creative's Sound Blaster. The High Court accepted that Aztech's dealing with the related software program constituted fair dealing for the purpose of private study and was therefore a valid defense.

Creative appealed against the High Court's decision, and in *Creative Technology Ltd* v. *Aztech Systems Pte Ltd (1997) 1 SLR 621*, the Court of Appeal reversed the finding of the High Court. The Court of Appeal found that there was disassembly by Aztech, which involved a degree of reproduction and adaptation which revealed the ideas and interfaces of Creative's program.

It has been suggested that it is probable that the copying and memory referred to in the above case is copying to the random access memory (RAM) of the computer.[31] This would mean that the reference in Section 17 of the Copyright Act to storage in a computer also includes temporary storage in the random access memory of a computer.[32]

(II) Authorizing Infringement

Under authorizing infringement, a person who authorizes another to commit primary copyright infringement is liable to the copyright owner. It should be noted that authorizing infringement will be found only if primary infringement also exists. For example, in the Singapore case of *John Robert Powers School Inc* v. *Tessensohn t/a Clea Professional Image Consultants (1993) 3 SLR 724,* the plaintiff claimed that the defendant's employees photocopied copyrighted notes and the defendant, by making the photocopies freely available, was deemed to have authorized the employees' copyright infringement. The High Court, however, did not accept the plaintiff's argument as there was no evidence of any photocopying of the notes by the defendant's employees or students. The Court's finding, therefore, was that, unless there is an act of primary infringement, there can be no authorizing infringement.[33]

What amounts to authorizing an infringement may be gleaned from the Court's decision in another Singapore case: *Ong Seow Pheng and Others* v. *Lotus Development Corp and Others (1997) 3 SLR 137.* In this case the appellants were found in possession of unlicensed copies of the respondent Lotus' software programs and manuals. It was held by a judge-in-chambers that the appellants: (a) authorized the retailers (including one Lur, a software pirate) to infringe the copyright in the whole of the respondents' software package; (b) were liable as joint tortfeasors with the retailers as they were acting pursuant to a common design, namely, to infringe the respondents' copyright.

However, on appeal, the Court of Appeal held that the appellants might have facilitated and even incited Lur's infringements, but this was not the same thing as authorization. Once the appellants sold and delivered the infringing copies of the manuals or programs to Lur, these copies were out of their hands and they had no control over who knew that the appellants were in no position to grant the right to make infringing copies. On the evidence, the Court of Appeal was also unable to accept Lotus' submission that the appellants authorized Lur and other retailers to make copies of the software program sold to them.

(III) Secondary Infringement

Under secondary infringement, a person who trades in an article with the knowledge that it is an infringing article is liable to the copyright owner.[34] Apart from the right of the copyright owner to claim remedies against such an infringer, the infringer may also face criminal liability under Section 136 of the Copyright Act.

Just as an action for authorizing infringement requires there to be primary infringement in the first place, secondary infringement similarly requires the articles in question to be infringing copies of the copyrighted work. An action for secondary infringement fails if the articles are genuine copies. For instance, in the Singapore case of *Television Broadcasts Ltd* v. *Golden Line Video & Marketing Pte Ltd (1988) SLR 930*, the copyright owner and its exclusive licensees (the plaintiffs) sued the defendant for secondary infringement as the defendant had rented out video tapes of the copyrighted movies without license from the copyright owner. Instead, the defendant had obtained copies of the movies that were made by the exclusive licensee in Singapore. The High Court rejected the plaintiffs' claims and struck out the action, explaining that the Copyright Act did not give the copyright owner exclusive rights of distribution in relation to the film. Rather, the right of distribution was in relation to infringing copies of the film.[35] As the copies of the film rented out by the defendants in this case were noninfringing tapes, the action by the plaintiffs could not be sustained.

Remedies for Copyright Infringement

A copyright owner whose rights have been infringed may commence a copyright infringement action under the Copyright Act[36] in the subordinate courts. The copyright owner may expect one of the following remedies to be granted by the Court:[37]

(a) an injunction (subject to such terms, if any, as the Court thinks fit);
(b) damages;
(c) an account of profits; or
(d) statutory damages.

Statutory damages may be awarded if elected by the copyright owner in lieu of damages or an account of profits. Under Section 119(2)(d) of the Copyright Act, the amount of statutory damages to be awarded shall be:

(i) not more than S$10,000 for each work or subject matter in respect of which the copyright has been infringed; but
(ii) not more than S$200,000 in the aggregate, unless the copyright owner proves that his actual loss from such infringement exceeds S$200,000.

Infringement of Layout-Designs

Like proprietors of other types of intellectual property such as patent and copyright, as discussed in the above sections of this chapter, a qualified owner of a layout-design is protected under the Layout-Designs of Integrated Circuits Act (Cap 159A) and he is also entitled to exploit his layout-design to the exclusion of others.[38]

Therefore, by way of illustration, if Mr. Y is a qualified owner of a layout-design, Mr. X exploiting the layout-design without Mr. Y's knowledge or permission would amount to an infringement of Mr. Y's rights granted under the Layout-Designs of Integrated Circuits Act.

This is pursuant to Section 9 of the Layout-Designs of Integrated Circuits Act, which states that it is an infringement of a qualified owner's right in a protected layout-design for any person to copy or commercially exploit the layout-design without the consent of the qualified owner.

Remedies for Layout-Design Infringement

Where a layout-design has been infringed, the qualified owner may seek legal recourse by commencing legal proceedings against the infringer. Section 12 of the Layout-Designs of Integrated Circuits Act sets out the remedies which are available to the qualified owner. It provides that a qualified owner may take whatever proceedings and seek whatever remedies by way of damages, injunctions, accounts, or otherwise with respect to an infringement of his rights in a protected layout-design, as are available with respect to any other property rights.

Therefore, it may be gleaned from the above provision that remedies available to the qualified owner include:

(a) applying for an injunction to stop the infringer from infringing upon the layout-design rights. This is a powerful remedy available to the qualified owner of the layout-design which has been infringed, and it involves applying to the Court to obtain an injunction to prevent any future infringement by the infringer.

There are two types of injunctions. The first one is an interim injunction which is granted very early in court proceedings. This will restrain the defendant from infringing the layout-design pending a resolution by the Court (or unless parties come to an out-of-court settlement). The second type of injunction is a final injunction, which is usually granted by the Court after it finds that there has been an infringement. In effect, it simply means that the interim injunction has now become permanent.

To obtain an injunction, the plaintiff will have to establish that:

(i) there is a serious question to be tried;
(ii) if the plaintiff were to succeed at trial, damages would not be adequate compensation for his loss;
(iii) the balance of convenience (for not granting an injunction) lies in the plaintiff's favor; and
(iv) there are special circumstances in the plaintiff's favor (or alternatively, there are no special circumstances that are in the defendant's favor).

While an injunction is in place (regardless of whether it is an interim or final injunction) and if the defendant violates the terms of the injunction, the plaintiff will be at liberty to commence contempt of court proceedings. It is a serious matter if the defendant is found guilty of contempt of court. The defendant can be fined and/or imprisoned. Usually, once an injunction is in place, the defendant will rarely risk violating the terms of the injunction order.

(b) damages. This is the monetary award that will be granted by the Court to compensate the qualified owner for the loss that he has suffered as a result of the layout-design infringement.

Usually after the trial, separate proceedings will be commenced by the qualified owner to assess the damages that he is to be awarded. Unfortunately, the assessment of damages may also require the assistance of experts and may be a time-consuming exercise. The Layout-Designs of Integrated Circuits Act also allows the Court to award additional damages, should the Court feel that it is proper after taking into consideration all the circumstances of the case, including but not limited to the *flagrancy of the infringement* and also the benefits the infringer has obtained as a result of the infringement, that is, the profits the infringer made;[39] and

(c) account of profits. This is different from damages and requires a detailed review of the profits made by the infringer as a result of infringement, and whatever gains the infringer has made as a result of the infringement. Any profits or gains must be surrendered to the qualified owner.

However, this remedy may not be popular due to the difficulty in determining the profits that arose out of the infringement. It could be a costly exercise to assess the extent to which the infringer's profits had been increased by the infringement. It may also be necessary to determine how the total profit is to be apportioned between the infringer's legitimate business and the increase in profit that resulted from the infringement.[40]

Additional Remedies: Order for Delivery-Up or Destruction of Infringing Layout-Design

The Layout-Designs of Integrated Circuits Act also provides for additional remedies to protect the layout-design. First, the qualified owner may, under Section 13 of the Layout-Designs of Integrated Circuits Act, apply to the Court for an order that the infringing article be delivered to him or to another person that the Court specifies.[41] Second, pursuant to Section 14 of the Layout-Designs of Integrated Circuits Act, the qualified owner may also apply to the Court for an order that the infringing article delivered up be forfeited, destroyed, or disposed of as the Court specifies.[42] In each case, the Court shall have regard to whether other remedies are available to the qualified owner and whether such remedies would be adequate to compensate him and protect his interests.[43]

Thus, by way of an illustration, a qualified owner Mr. X whose layout-design has been infringed by Mr. Y, apart from an action for damages and/or an account of profits, Mr. X may also apply to the Court for Mr. Y to deliver up the infringing layout-design in Mr. Y's possession, and for the layout-design to be subsequently disposed of or destroyed so that no further infringements can be made.

However, the Court will order the disposal of the infringing layout-design only after having regard to whether the other remedies available to the qualified owner would adequately compensate and protect his interests. Furthermore, an order for disposal shall only take effect after the time allowed for appeal has passed, or until final determination of the court proceedings, or until the court proceedings are abandoned on appeal.[44] This demonstrates that the Court does not easily grant the remedy of disposal. Further, in light of what is stated in Section 14 of the Layout-Designs of Integrated Circuits Act, it is seen as a remedy of last resort.

Section 14(6) of the Layout-Designs of Integrated Circuits Act also states that if there is more than one person interested in the infringing layout-design that is delivered up, the Court may instead order that the offending layout-design be sold and the proceeds divided, or any other court order viewed as just in the circumstances.

Defenses

For a person alleged to have infringed upon a layout-design right, Section 10 of the Layout-Designs of Integrated Circuits Act sets out the various defenses to layout-design infringement. Thus, in the situation in which Mr. Y is alleged to have infringed upon Mr. X's layout-design rights, the defenses Mr. Y could rely upon may include situations whereby:

(a) the copying (of Mr. X's layout-design) is of any part of a protected layout-design that does not comply with the requirement of originality referred to in Section 5(1) of the Layout-Designs of Integrated Circuits Act. Hence, in brief, the layout-design that is allegedly copied is not *original* as required by the Layout-Designs of Integrated Circuits Act in order for rights to subsist. However, should a defendant wish to raise this defense, he will most likely need to engage the services of an expert having relevant experience and knowledge in the field of technology to explain to the Court why the layout-design is not *original;*

(b) the copying is done for a private purpose and not for the purpose of commercial exploitation. This would usually mean that the alleged infringer has not financially benefited from the copying;

(c) the copying is done for the sole purpose of evaluation, analysis, research, or teaching and to use the results of such evaluation, analysis or research to create a different layout-design that complies with the requirement of originality referred to in Section 5(1) of the Layout-Designs of Integrated Circuits Act;

(d) Mr. Y is himself a qualified owner of another protected layout-design that is (i) identical to Mr. X's layout-design; and (ii) independently created. Therefore, it is a defense if a person can show that he created an identical layout-design to another protected layout-design, and the identical layout-design was independently created. However, the chances of two persons creating identical layout-designs independently of each other would be relatively slim;

(e) The alleged infringer had the consent of the qualified owner. Thus, Mr. Y commercially exploits a copy of Mr. X's layout-design, or an integrated circuit in which Mr. X's layout-design is incorporated, or an article that contains an integrated circuit in which Mr. X's layout-design is incorporated, whether in Singapore or elsewhere, with the consent of Mr. X; or

(f) Mr. Y innocently infringes upon Mr. X's layout-design.[45] By virtue of Section 11 of the Layout-Designs of Integrated Circuits Act, an infringement is innocent if when the alleged infringer obtained a copy of the layout-design, integrated circuit or article, the person did not know or could not have reasonably been expected to know that the copy was unauthorized, or the integrated circuit was unauthorized, or that the article contained an unauthorized integrated circuit.

However, once the alleged infringer becomes aware, or could reasonably be expected to become aware that the copy or integrated circuit was unauthorized or the article contained an unauthorized integrated circuit, innocent infringement can only be relied upon for any subsequent commercial exploitation if the *alleged infringer* pays the qualified owner such remuneration as agreed, or determined by a method both parties agree to, or in the event that there is no agreement, either party applies to Court and the Court determines the remuneration.[46]

The test for whether a copy of a protected design is unauthorized, or if an integrated circuit which incorporates a protected layout-design is unauthorized, is whether the said copying or the incorporation was done without the consent of the qualified owner.

Reverse Engineering

This section intends to elaborate upon whether reverse engineering would amount to an infringement under the Layout-Designs of Integrated Circuits Act. Reverse engineering is defined as the process of analyzing intellectual property that is lawfully acquired with the intent of determining how the property operates.[47] The object of analysis is taken apart to see how it works in order to copy (or enhance) the product. Reverse engineering can be a concern for owners of layout-designs of integrated circuits, especially in light of the time and money that is spent on each layout-design.

However, the Layout-Designs of Integrated Circuits Act does not specifically prohibit reverse engineering. As stated above, Section 9 of the Layout-Designs of Integrated Circuits Act states that it is an infringement of a qualified owner's right in a protected layout-design to do any acts referred to in Section 8 of the Layout-Designs of Integrated Circuits Act (that is, to copy or commercially exploit) the layout-design without the consent of the qualified owner. Thus, the Layout-Designs of Integrated Circuits Act is silent on the issue of reverse engineering, even though the idea of finding out how an object works is usually executed to commercially exploit it.

The issue of reverse engineering under the Layout-Designs of Integrated Circuits Act has not been brought before the Courts of Singapore. It will be interesting to see how the Courts rule on the issue of reverse engineering. One approach would be to say that it is for the legislation to be amended and, until that time, the Court is likely to rule that reverse engineering is to be allowed.

Another approach would be to say that it is against public policy to allow reverse engineering. The Singapore Government has taken many active steps to promote Singapore as a vigilant and secure IP hub. In fact, Singapore has been voted the most IP-protected country in Asia for the last two years by the Political and Economic Research Consultancy.[48] As such, it can be argued that the Court is unlikely to sanction reverse engineering as it allows through the back door what cannot be done through the front door.

Infringement of Trademarks

The proprietor of a registered trademark has the exclusive rights to use the trademark and to authorize other persons to use the trademark, in relation to the goods or services for which the trademark is registered.[49]

Definition of Trademark Infringement

Under Singapore law, what amounts to an infringement of a trademark is set out in Section 27 of the Trade Marks Act (Cap 332). Under Section 27(1) of the Trade Marks Act, a person infringes a registered trademark, if without consent of the proprietor of the trademark he uses, in the course of trade, a sign that is identical to the trademark in relation to goods or services that are identical to those for which it is registered.

The infringer is presumed to cause a likelihood of confusion as a result of such use. A sign or mark is identical with a registered trademark only if it reproduces, without any modification, omission or addition, all the elements constituting the trademark or, if viewed as a whole, it contains differences so insignificant that they may go unnoticed by an average, reasonably well-informed, observant consumer.[50]

To better illustrate the above legal principle, the Singapore case of *Pan-West (Pte) Ltd* v *Grand Bigwin Pte Ltd (2003) 4 SLR 755*, will shed light on its application to the subject matter. In this case, the plaintiff (owner of the registered trademark) and the infringer were competitors in selling golf clubs. The plaintiff was the registered proprietor of the Singapore series mark *Katana Golf* since 1997, whereas the infringer was using the mark *KATANA* which was not registered. The Court held that the distinctive and dominant component of the trademark was the word *Katana*, and this had been reproduced in terms. In determining the identity of signs, the infringing signs did not have to be exactly the same in order to be considered identical. The plaintiff established a claim for trademark infringement under Section 27(1) of the Trade Marks Act.

Where the trademark is not identical to the registered trademark, Section 27(2) of the Trade Marks Act provides another set of definitions. To establish infringement of trademarks under Section 27(2) of the Trade Marks Act, it must be shown that the infringer has committed one of the following acts, and there exists a likelihood of confusion on the part of the public:

(a) the sign is identical with the trademark and is used in relation to goods or services similar to those for which the trademark is registered; or
(b) the sign is similar to the trademark and is used in relation to goods or services identical with or similar to those for which the trademark is registered.

The decided case law suggests that there are three aspects to similarity: visual, aural or phonetic, and conceptual. For example, in the case of *Pensonic Corporation Sdn Bhd v Matsushita Electric Industrial Co. Ltd (2008) SGIPOS 9*, the registered trademark of *PANASONIC* and the infringer's mark *PENSONIC* were found to be more similar than dissimilar as they are aurally and conceptually similar.

Noninfringement of Trademark

Section 28 of the Trade Marks Act sets out certain exceptions to trademark infringement, in particular, when the use of the trademark is in accordance with honest practices in industrial or commercial matters. These exceptions to trademark infringement are elaborated as follows:

(a) Use of a Person's Name

There would be no trademark infringement by a third party, if the third party uses his name or the name of his place of business, or the name of his predecessor in business or the name of his predecessor's place of business. This usually applies to a family business that has the family's name as the trademark, which is known to the public and the said business has been in the industry for a long period of time.

(b) Descriptive Use

There would be no trademark infringement if a third party uses a trademark to indicate the kind, quality, quantity, intended purpose, value, geographical origin or other characteristic of goods or services. There also will be no infringement if the person uses a trademark to indicate the time of production of goods or the time of rendering the services.

Another defense is when a third party uses the trademark to indicate the intended purpose of goods (in particular as accessories or spare parts) or services.

(c) Prior Use of Trademark

This defense allows the continuation of the use of an unregistered trademark by a third party who can prove that he has been using the trademark even before the date of registration of the other party's trademark, or the date the other party or his predecessor in title first used the trademark, whichever is earlier. This defense provides protection to a third party who is first to use a particular mark.

(d) Other Acts Not Amounting to Infringement of Trademark

A person who uses a registered trademark does not infringe the trademark if such use constitutes fair use in comparative commercial advertising or promotion, or the use is for a noncommercial purpose, or the purpose of news reporting or news commentary.

Revocation and Invalidation of a Trademark

Registration of a trademark, on the face of it, grants the proprietor of the trademark the rights to use the mark. However, any third party who has a valid reason may apply to revoke or invalidate the trademark from the Trade Mark Register.

(a) Revocation

There are several grounds for revocation of a trademark and they are set out in Section 22 of the Trade Marks Act. A trademark can be revoked if any one of the following grounds is established:

(i) if within the period of 5 years after the date of completion of the registration procedure, the mark has not been put to genuine use by the proprietor or there are no proper reasons for nonuse; or
(ii) the use of the mark has been suspended for an uninterrupted period of five years and there are no proper reasons for nonuse; or
(iii) in consequence of acts or inactivity of the proprietor, the mark has become the common name in the trade for the products or services for which it is registered; or
(iv) in consequence of the use made of it by the proprietor or with his consent, it is liable to mislead the public, particularly as to the nature, quality or geographical origin of those goods or services.

(b) Invalidation

There are several grounds for invalidation of a trademark, and they are set out in Section 23 of the Trade Marks Act. A third party may apply to invalidate a trademark based on any one of the following grounds:

(i) if the trademark was registered in breach of Section 7 of the Trade Marks Act, for example, the mark does not qualify as a trademark or it lacks distinctiveness, or the application for registration was made in bad faith; or
(ii) if the mark conflicts with an earlier trademark or an earlier right;[51] or
(iii) on the grounds of fraud in its registration, or that the registration was obtained by misrepresentation.

Remedies for Trademark Infringement

When a trademark has been infringed within the meaning of Section 27 of the Act as explained above, the trademark owner may seek legal recourse by commencing legal proceedings against the infringer in the High Court.

The remedies that the High Court may order in relation to trademark infringement include any one of the following:[52]

(a) an injunction (subject to such terms, if any, as the Court thinks fit);
(b) a monetary award (damages; account of profits; statutory damages);
(c) an order for erasure of the offending sign; and/or
(d) an order for delivery up and disposal of the infringing goods.

Damages and Monetary Award

The damages awarded for trademark infringement are normally in the form of loss of profit suffered by the trademark proprietor as a result of the trademark infringement. The loss of profit must be established by the proprietor (the plaintiff) and it cannot be assumed that every sale that the infringer (the defendant) makes of the infringing goods would be a sale that the proprietor would have made.[53] In cases where there are difficulties proving the loss of profit, the Court may endeavor to form a *rough estimate* of the loss of profit after reviewing all the circumstances and evidence.[54]

Pursuant to Section 31(3) of the Trade Marks Act, when the Court awards any damages, the Court may also make an order for an account of any profits attributable to the infringement that have not been taken into account in computing the damages.

In any action for infringement of a registered trademark where the infringement involves the use of a counterfeit trademark in relation to goods or services, the plaintiff is entitled, at his election, to:

(a) damages and an account of any profits attributable to the infringement that have not been taken into account in computing the damages;
(b) an accounts of profits; or
(c) statutory damages[55] (not exceeding S$100,000 for each type of goods or service in relation to which the counterfeit trademark has been used and not exceeding in the aggregate S$1 million, unless there is proof that actual loss from such infringement exceeds S$1 million).[56]

Where the infringer is found to have infringed a registered trademark, the Court may make an order requiring the infringer to cause the offending sign to be erased, removed or obliterated from any infringing goods, material or articles in his possession, custody or control.[57] If it is not reasonably practicable for the offending sign to be erased, removed, or obliterated, the Court may make an order to secure the destruction of the infringing goods, material or articles in question.[58]

If a person found to have infringed a registered trademark has in his possession, custody or control any counterfeit goods, the Court shall order that the counterfeit goods be delivered up to such a person as the Court may direct for destruction. However, this is only available if the plaintiff applies to the Court for such an order and there are no exceptional circumstances which justify the refusal of such an order.[59]

Costs

A registered trademark owner can also claim against the infringer for the costs of commencing proceedings in the High Court, pursuant to Order 59 of the Rules of Court.

Remedy for Groundless Threats of Infringement Proceedings

In balancing the rights of the trademark proprietor and the alleged infringer, the law also provides remedies for a person who is aggrieved by a groundless threat of infringement proceedings made by any person including, but not limited to, the trademark proprietor. The aggrieved person can seek a declaration that the threats are unjustifiable, an injunction against continuance of the threats, and damages in respect of any loss suffered by the threats.[60]

However, this remedy available to the alleged infringer cannot be invoked if the threat by the plaintiff relates to:

(a) the application of the mark to goods or to material used or intended to be used for labeling or packaging goods;
(b) the importation of goods to which, or to the packaging of which, the mark has been applied; or
(c) the supply of services.[61]

Avoid Infringement and Measures for Safekeeping of Intellectual Property

After discussing the various types of infringement of intellectual property rights applicable to integrated circuits, the following short section seeks to identify ways of how intellectual property infringement could be avoided and how a person should safe-keep his intellectual property, since taking steps to protect one's intellectual property rights also involves wariness or cautiousness to avoid infringing another person's intellectual property rights.

It is always prudent that detailed documentary records are being kept when a person is creating or developing his work. This practice ensures that when a third party alleges any IP infringement, the IP owner is able to adduce evidence to prove a paper trail demonstrating the originality of his work, and that he is indeed the proper owner of such IP. Examples of such documentary evidence are inventor's notes, progress reports, results of any test runs/experiments, etc. As more of these records are documented in computer softcopy, one should always ensure that backup softcopies of these records are also made.

Furthermore, if during the development of the original intellectual property, it becomes necessary for the owner of the intellectual property to share his preliminary creations with a third party (for investment purposes, to verify his invention, to seek feedback, etc), it is also good practice to enter into a nondisclosure agreement or confidentiality agreement with the third party. By so doing, the owner of the intellectual property safeguards his IP by way of a contract and the third party is also prevented from using or sharing whatever information he receives from the owner of the IP for his own benefit. In the event of a breach of the nondisclosure agreement, the owner of the IP could seek legal recourse by commencing legal proceedings against the recipient of such information.

In the event that intellectual property is owned by a company or a body corporate, then it is prudent for the company to enter into contracts with its employees, agents, and independent contractors to ensure that it is emphasized and being made clear to them that the Company is the lawful owner of its intellectual property and that such IP cannot be used without the company's consent.

To ensure that a company's employees, agents, or independent contractors do not infringe a third party's intellectual property rights, it is recommended that the parties involved should sign contracts with the company, acknowledging the company as owner of such intellectual property and to observe the company's rules and regulations when using such intellectual property.

It is also recommended that regular IP seminars and talks be conducted for the company's employees. These seminars serve the purpose of educating the company's staff on the importance of intellectual property rights and the consequences of infringing those intellectual property rights. These regular seminars could also promote IP awareness within the company and at times improve the company's innovations.

Summary

Patent Infringement

Ownership of a patent entitles the owner to exercise a monopoly over the patent. Any other party who commits a prohibited act in relation to the patent may be sued by the patent owner for patent infringement. Patent infringement would be successfully made out if (a) the act committed by the other party (defendant) is a prohibited act within the meaning of Section 66 of the Patents Act; and (b) the infringing product or process falls within the claim of the patent. On the other hand, a defendant who successfully proves in his defense that the act falls within Section 66(2)(a)-(i) of the Patents Act would not be liable for patent infringement as acts prescribed therein constitute defenses to any claim of patent infringement.

To remedy a patent infringement, a patent owner may seek legal remedies available under Section 67 of the Patents Act, including an injunction to restrain further infringement, an order to deliver up and destroy the infringing product or process, damages, or account of profits. In relation to damages and account of profits, it should be noted that according to Section 67(2) of the Patents Act, a patent owner may only elect either damages or an account of profits, but not both.

Copyright Infringement

A copyright owner is similarly entitled to exclusive rights in literary works, including computer programs, set out in Section 26(1)(a) of the Copyright Act. Copyright infringement can take three forms: (i) primary infringement; (ii) authorizing primary infringement; and (iii) secondary infringement.

Primary infringement occurs when a person does in Singapore any unauthorized act in the copyright without the license of the copyright owner. Primary infringement is successfully made out if (i) the infringing act was committed in Singapore; (ii) the infringer copied the work of the copyright owner; and (iii) the infringement is of a substantial part of the copyright work.

Copyright infringement under authorizing infringement occurs when a person authorizes another person to commit primary infringement of a copyright. Thus, an authorizing infringement will be found only if primary infringement exists. In *Ong Seow Pheng and Others* v. *Lotus Development Corp and Others (1997) 3 SLR 137*, the Court held that authorization must be distinguished from facilitation or incitement. Unlike authorization of infringement, facilitation or incitement will not result in liability for copyright infringement.

Third, secondary infringement occurs when an infringer becomes liable if he trades in an article with the knowledge that the article infringes the copyright owner's work. As with authorizing infringement, secondary infringement will be found only if primary infringement exists. In addition, the infringer may also face criminal liability under Section 136 of the Copyright Act.

As remedies for a copyright owner whose copyrights have been infringed, the Court may grant an injunction, damages, an account of profits, or statutory damages under Section 119(2)(d) of the Copyright Act.

Layout-Design Infringement

In brief, layout-design infringement occurs when there has been copying of a protected layout-design of an integrated circuit that is being commercially exploited without the consent of its qualified owner.

To identify whether a protected layout-design or integrated circuit has been infringed it is prudent to first identify what has been infringed. For this, it will be necessary to engage the services of an expert having relevant experience and knowledge in the field of technology. Once the expert has confirmed that there has been an infringement, the qualified owner will be able to commence legal proceedings against the infringer. At the start of the proceedings, the qualified owner, to maintain the status quo, can apply to Court and obtain an interim injunction (which will be made into a final injunction if the owner succeeds at trial).

The Layout-Designs of Integrated Circuits Act also provides the defenses that are available to the alleged infringer. One of the defenses is that the layout-design that is copied is not an original creation and thus, it is not protected as a layout-design as defined by the Layout-Designs of Integrated Circuits Act. Other defenses that can be raised include that the copying was done for a private purpose or for the sole purpose of evaluation, analysis, research, or teaching. It is also possible to raise the issue of innocent infringement. However, if a person is raising this as a defense, he must pay remuneration to the qualified owner once it is known (or could have reasonably been known) that infringement has taken place.

Once the qualified owner succeeds in his infringement proceeding, the qualified owner can obtain a final injunction and damages or an account of profits. In certain situations (say for example, if the defendant flagrantly infringes the protected layout-design), the Court may be inclined to award additional damages in favor of the qualified owner.

Trademark Infringement

The proprietor of a trademark has exclusive rights to use, and to authorize others to use, the trademark in relation to the goods or services for which the trademark is registered.

To use in the course of trade, a sign which is identical to the trademark in relation to goods or services which are identical with those for which it is registered would amount to an infringement under Section 27(1) of the Trade Marks Act. A sign or mark is considered identical to a registered trademark if it reproduces, without any modification, omission or addition, all the elements that make up the trademark, or if the differences between the two marks are so slight that they cannot be observed by the average consumer.

Where a sign is not identical to the registered trademark, Section 27(2) of the Trade Marks Act states that to establish infringement under the section it must be shown that (a) the sign is identical to the trademark and is used in relation to goods or services similar to those for which the trademark is registered; or (b) the sign is similar to the trademark and is used in relation to goods or services identical with or similar to those for which the trademark is registered. Further, it must also be shown that for either (a) or (b), there exists a likelihood of confusion on the part of the public.

The defendant in a trademark infringement case may rely on several defenses to refute the claim. These are set out in Sections 28 and 29 of the Trade Marks Act and include use of a person's name in a trademark, descriptive use of a trademark, and prior use of a trademark.

Where a trademark has been infringed within the meaning of Section 27 of the Trade Marks Act, the proprietor may obtain remedies from the Court in the form of an injunction, a monetary award, an order for the erasure of the offending sign and/or an order for delivery up and disposal of the infringing goods. The proprietor may also claim against the infringer for costs of commencing proceedings in the High Court, pursuant to Order 59 of the Rules of Court.

The law also protects a person who is aggrieved by a groundless threat of infringement proceedings made by any person (including but not limited to the proprietor of the trademark). Under Section 35 of the Trade Marks Act, the aggrieved person may seek a declaration that the threats are unjustifiable, an injunction against the continuation of such threats, and damages in respect of any loss suffered by him as a result of such threats. However, Section 35 of the Trade Marks Act cannot be relied upon where the threat is in relation to (a) the application of the mark to goods or to material used or intended to be used for labeling or packaging goods; (b) the importation of goods to which, or to the packaging of which, the mark has been applied; or (c) the supply of services.

While registration of a trademark grants the proprietor of the trademark exclusive monopoly rights, such rights may be challenged by a third party who, with a valid reason, applies to revoke or invalidate the trademark from the Trade Mark Register. Grounds for revocation of a trademark are set out in Section 22. These are (i) the mark has not been put to genuine use by the proprietor or there are no proper reasons for nonuse within five years after the date of completion of registration procedure; (ii) use of the mark has been suspended for five or more uninterrupted years and no proper reasons exist for their nonuse; (iii) the mark has become a common name in the trade for which it was registered due to the proprietor's inactivity; or (iv) the mark misleads the public as to the nature, quality, or origin of the goods and services for which the mark was registered. Grounds for invalidation are prescribed under Section 23 of the Trade Marks Act and include (i) failure of the mark to qualify as a trademark as it lacks distinctiveness, or was registered in bad faith; (ii) conflict between the mark and an earlier trademark or an earlier right; or (iii) obtaining the registration as a result of fraud or misrepresentation.

7

Procedures and Formalities for Protection of Intellectual Property Rights Applicable to Integrated Circuits

Introduction

The previous chapters discuss in length the various legal principles and concepts in relation to intellectual property rights applicable to integrated circuits. For integrated circuits to be protected by such intellectual property rights, several of them require formal registration with the authority in order for these intellectual property rights to be enforceable against infringement.

As highlighted in Chapter 5, the four main types of intellectual property rights that are applicable to integrated circuits are patent, copyright, layout-design of integrated circuits, and trademark. To obtain patent and trademark protection for integrated circuits, the proprietor is required to undergo a formal application or registration and a renewal process through the Registry of Patents and the Registry of Trade Marks, respectively. This application or registration process, together with the renewal process (when the patent or trademark is eventually granted by the authority), are complex processes involving many procedural steps, formalities, and documentation requirements. Often, to coordinate such formal application or registration with the registries, the proprietor engages qualified patent agents or trademark agents to assist in the entire application or registration process.

It is deliberate that only patent application and trademark registration processes are highlighted for discussion. The earlier chapters emphasize that for legal protection offered by copyright or layout-designs of integrated circuits, there is no formal registration requirement or process with the authority.

Copyright

It has been pointed out several times in the earlier chapters that there is no requirement to formally register copyright with any authority or governmental body in Singapore in order to seek legal protection or to commence a legal action against an infringing party for copyright infringement related to integrated circuits. Copyright subsists automatically in copyright works when these original works are created or published and the duration of copyright protection varies depending on the types of copyright works. More details are set out in Chapter 5 of this book.

Layout-Designs of Integrated Circuits

Similar to copyright protection, the legal protection for integrated circuits provided by the Layout-Designs of Integrated Circuits Act (Cap 159 A) requires no formal registration with any authority. Protection of the layout-design of an integrated circuit commences when it is first created. The duration of the protection offered is explained in greater detail in Chapter 5.

Filing a Patent Application in Singapore

This section provides a basic description of the patent application process in Singapore after an initial patent application submission. In brief, the patent application process can be subcategorized into five major steps consisting of:

- Preliminary Examination
- Search and Examination
- Publication
- Grant of Patent
- Renewal of Patent

Preliminary Examination

Upon the submission of a patent application, the Registry of Patents, Intellectual Property Office of Singapore (IPOS), proceeds to conduct a preliminary examination to ensure that all forms and documents submitted fulfill the requirements of the Patents Act and Rules. A complete patent application consists of:

(a) a request for grant of a patent
(b) specification detailing the descriptions of the invention
(c) one or more claims
(d) drawings of the invention if applicable
(e) the abstract[1]

The Singapore Patent System allows for the non-filing of claims at the point of filing the application. The applicant is given a grace period of 12 months from the date of filing of the application to file the claims in the event that no priority date is claimed. However, if the application claimed priority to an earlier application, the applicant is required to submit the claims 12 months from the priority date or two months from the date of filing, whichever is later.[2]

If the patent application is in order and the prescribed official fee is paid, an application number and date is assigned by the Registry of Patents.[3] From the point of issuance of the application number and filing date, the invention is categorized as *patent pending* until the patent is granted. The extremely crucial date of filing is derived from the precise date when the patent application is submitted to the Registry of Patents and acts as a priority date in the absence of earlier filings, and also when the applicant decides to file further corresponding applications in different jurisdictions.

Once the application has been accepted by the Registry of Patents, a clear formalities report is issued and the applicant is then able to prosecute the application further within a specific time frame as indicated by the Patents Act and Rules. Such a time period is typically calculated from the priority date of the application or, in the absence of a priority date, the date of filing is used.

Publication

The patent application is published 18 months from the date of filing or the priority date, whichever is earlier.[4] The bibliography of the application, together with the abstract and a suitable drawing, if any, is reflected in the monthly Singapore Patents Journal. Upon publication, the patent application is now open for public inspection. The concept of *laid open* implies that all the documents pertaining to the application are made available to the general public. It should be noted that even after the patent application is made public, the applicant still reserves his legal rights to the patent application. However, in the case wherein the applicant does not wish to disclose his invention, the applicant should proceed to withdraw the patent application from the patent office before the 18 month publication deadline.

Search and Examination

The next stage of the patent application process involves making a request for Search and Examination. Search and Examination is conducted when the applicant formally files a request within the time frame and pays the prescribed official fee. The Singapore patents system offers a vast degree of flexibility at this stage of the patent application process. The present Singapore patent system provides a two-track application system for the prosecution of the patent application. The applicant may elect to proceed under the default fast-track system, or he may elect the slow-track system. The applicant, depending on his patenting strategy, may:

(a) First file a Request for Search (within 13 months from the priority date or date of filing, whichever is earlier) before proceeding to file a Request for an Examination (within 21 months from the priority date or date of filing, whichever is earlier) upon receipt of the Search Report; or
(b) Ensue to request, within 21 months from the priority date or the date of filing, whichever is earlier, for a combined Search and Examination

In the event that the applicant has proceeded to file corresponding applications into prescribed countries, or a corresponding international application filed under the Patent Cooperation Treaty (PCT), options (c) and (d) below may prove to be the most cost effective to further prosecute the Singapore application:

(c) By relying on a Search Report issued by one of the search authorities of the prescribed countries, or the PCT for a corresponding patent application, the applicant may proceed to only file a Request for Examination within 21 months from the priority date or the date of filing, whichever is earlier. A corresponding application refers to an application filed into the patent office of a prescribed country or the PCT sharing the same priority claims as the Singapore application.[5] The prescribed countries in this instance include the United States, the European Patent Office (for applications filed in English), United Kingdom, Australia, Canada (for applications filed in English), Japan, Republic of Korea, and New Zealand.[6] This route provides a cost-saving measure wherein the search fee is exempted.

(d) Alternatively, the applicant may, within 42 months from the priority date or the date of filing, whichever is earlier, file the prescribed information issued by the patent offices of the prescribed countries or the PCT, indicating the allowance or grant of a corresponding application. The prescribed information includes either:

  (i) A copy of the granted patent, certified by the patent office of the prescribed country
  (ii) The final result of the Search and Examination as to substance, and a copy of the patent claims referred to in the final results.[7]

By employing the corresponding applications that have been filed and examined by an established patent office, duplication of work may be avoided, and a considerable amount of money and effort may be saved by the applicant in the course of patent prosecution. The availability of the above steps allows for a quick grant of the patent application and undeniably help the applicants in reducing their cost and effort in obtaining a patent in Singapore.

It is noteworthy that both the Search and Examination are not carried out by IPOS. Instead, the Search and Examination are carried out by three designated patent offices, (1) the Australian Patent Office, (2) the Austrian Patent Office, and (3) the Danish Patent Office.

When a Request for Examination is made, the examiner ascertains whether the invention meets the patentability requirements of novelty, inventive step, and industrial applicability. If the patent application is satisfactory and fulfils all the prerequisites, a favorable Examination Report is issued by the examiner. The patent application may then proceed to the next stage that involves payment for the grant of the patent. However, if the examiner establishes that the patent application does not comply with the patentability requirements, a written opinion is issued wherein the examiner raises any adverse matters.

The applicant is given an opportunity to respond to the examiner's written opinion. In short, the applicant is invited to make amendments or to argue against the examiner's objections. The applicant is given a five-month nonextendible period to consider the objections and to formulate a response in reply to the written opinion raised by the examiner.[8] The applicant may choose to respond to the written opinion. In the event that the applicant decides to rebut the opinions of the examiner, a written submission, together with the amendments sought, is filed. Filing a response to the written opinion initiates another cycle of examination wherein the examiner decides if the new submission is adequate to overcome his previous objections, or if a further written opinion is necessary.

In light of the above, the applicant may also choose not to respond to the written opinion issued by the examiner. Failing to file a response within the stipulated time limit results in the establishment of a final Examination Report, which reflects the objections cited in the written opinion.

Grant of Patent

The final stage of the patent application is the payment of the prescribed fee for the grant of the patent, which is issued as a Certificate of Grant. Prior to the issuance of the Certificate of Grant, the Registry of Patents will ensure that all of the following conditions have been satisfied by the applicant:

(a) all formal requirements have been complied with
(b) the following documents have been submitted to the Registry:

   (i) a favorable Search and Examination Report, stating that all the claims are novel, have inventive steps, and have industrial applicability
   (ii) a favorable Search and Examination Report from a patent office of a prescribed country indicating that a corresponding patent application which has been filed is allowed and is proceeding to grant
   (iii) a corresponding patent application which has been already granted in any one of the prescribed countries or
   (iv) an International Preliminary Report on Patentability of a corresponding application filed with the Patent Cooperation Treaty[9]

It should be noted that in the instance in which the applicant is paying the fee for grant through (ii), (iii), or (iv) set out above, the Singapore patent application should be amended to be consistent with the said corresponding patent application, which has been allowed or granted in any of the prescribed countries:

(c) Filing of a consolidated specification incorporating all amendments made in the course of the prosecution, if any
(d) Filing a prescribed form and paying the Fee for Grant

On the other hand, the applicant may switch to the slow-track option in the event that he has good reasons to do so. For example, the applicant may need more time to source for funding from venture capitalists for his invention, or the applicant wishes to elect option b(iii) or b(iv) set out above but the required documents for the corresponding applications have not been issued. The applicant may file a request for a block extension of time within 39 months from the priority date or the date of filing, whichever is earlier. By filing a request for a block extension of time, the deadline for prosecution of the patent application is delayed:

(a) When the Request for Search is filed earlier and the Request for Examination is not filed in 21 months, a request for examination can be filed within 39 months
(b) When the request for a combined Search and Examination is not filed in 21 months, the same can be filed within 39 months
(c) When the Request for Examination based on the search results of the corresponding application is not filed in 21 months, the same can be filed within 39 months
(d) When the prescribed information is not filed in 42 months, the same can be filed within 60 months

The Singapore patents system is a *self-assessment system;* that is, the patent grant will not be refused on the basis that the claims lack novelty, inventive step, or industrial applicability. The onus is on the applicant to ensure that the invention is patentable. The applicant bears the responsibility of evaluating the patent application, and he has to decide if it is worthwhile to maintain the said patent. Upon receiving the request for grant, the patent office proceeds to issue the Certificate of Grant and to publish the Grant of the Patent in the Patent Journal.

Renewal of Patent

The issuance of the patent gives the proprietor of the invention an exclusive monopoly to allow him to make, dispose of, offer to dispose of, use, import, or keep the invention either for disposal or otherwise. Nonetheless, this is only achievable if the applicant continues to pay the renewal fee for the patent. A granted patent can only be sustained if an annual renewal fee is paid to the patent office, usually for a term of 20 years from the date of filing, though it may be extended in certain circumstances (for example, delay in the process of obtaining marketing approval for a pharmaceutical product). Nonpayment of renewal fees will cause the patent to lapse. After the term of 20 years, the patent is deemed to expire, and the technology is open to the exploitation of the general public.

Filing a Patent Outside of Singapore

Prior to filing an application abroad, it is vital for the applicant or any one of the inventors, who are resident in Singapore, to obtain written permission from the Registry of Patents confirming that the invention does not breach any national security considerations.[10] Security clearance must be sought from the Registry of Patents in the event that:

(a) the overseas filing is the first patent application for the invention or
(b) an overseas application is to be filed within two months from the Singapore application

After two months from the filing of the Singapore application, the applicant may proceed to file the invention abroad, as it is assumed that the application has been examined and identified to lack national security concerns by the Patent Registrar.[11]

Filing an International Patent Application

Patent protection involves complex procedural and formality requirements when making patent applications to the relevant patent authorities. A cost-effective measure to ensure the broadest possible patent protection for an invention across various regions around the world is by filing an international patent application through the Patent Cooperation Treaty (*PCT*). The PCT is an international treaty administered by the International Bureau (*IB*) of the World Intellectual Property Office (*WIPO*) based in Geneva. This makes it possible for the applicant to file patent applications for an invention simultaneously in different countries with a single office, in one language, and using a single set of forms and fees. The PCT Contracting States[12] include a large number of countries spanning across most of the continents. However, as is further discussed in this section, although filing an international patent application merely serves as a means to secure patent protection in member countries, it does not provide for the grant of an internationally effective patent.

A PCT application may be filed by a person who is a national or resident of a Contracting State.[13] It may generally be filed with the national patent office of the Contracting State in which the applicant is a national or resident. Alternatively, upon the applicant's choice, a PCT application may also be filed directly with the IB of WIPO. In Singapore, the Intellectual Property Office of Singapore (*IPOS*) acts as the Receiving Office for all PCT applications.

The main procedures for filing an international patent application with the PCT can be outlined as follows:

- Filing an international patent application
- International publication
- International search and examination
- National phase

As previously discussed, an international patent application may be filed with the national Receiving Office (RO) of a contracting state, or directly with the IB of WIPO by a national of a PCT contracting state.[14] In any event, if there are numerous applicants for an invention, only one of the said applicants needs to be a national of a contracting state to permit the filing of an international patent application.[15]

An international patent application number is allocated, together with an international filing date—the date when the patent application is filed with the RO or IB,[16] once the patent application complies with the formal requirements established by the PCT. The applicant is given a grace period of one month from the international filing date to pay the prescribed official fees which cover the transmittal fee, international filing fee, and search fee.[17] The fees to be paid may vary depending on the International Searching Authority (ISA) elected by the applicant.

For a patent application filed with IPOS as a RO, the competent international searching authorities are as follows:

(a) Australian Patent Office
(b) Austrian Patent Office
(c) European Patent Office
(d) Korean Patent Office

International Publication

A patent application filed via the PCT is published for public inspection 18 months from the earliest declared priority date, if there is a priority claim or an international filing date.[18] However, it should be made clear that no third party is allowed to view the patent application before the expiration of the 18 month deadline.

Therefore, if the applicant does not wish for his patent application to be published, he may proceed to withdraw the patent application before the 18 month publication deadline expires.

International Search and Examination

An international search is conducted by the chosen ISA wherein the examiner looks for disclosures, patents, patent applications, or technical publications which relate to the field of the patent application being filed. This search is conducted using various databases of common languages in which patent applications are generally filed, including English, Chinese, Japanese, German, to name a few.

An International Search Report (ISR) and the written opinion of the ISA is established once the search is completed.[19] The ISR details the list of patent documents or applications cited to possibly affect the patentability of the patent application in question. These references are further categorized to indicate the degree of relevance to the filed international patent application. In the written opinion of the ISA, the examiner may raise objections with regard to the patentability of the invention. This provides an opportunity for the applicant to review the patentability of his invention.

The applicant is given an opportunity to amend only the claims of the international application.[20] These amendments should be filed directly with the IB within the time limit as stated by the PCT Regulations. The amendments must be submitted with a letter indicating the differences between the claims as filed and the claims as amended. The amendments may be accompanied by a statement explaining them, and the amendments shall not go beyond the disclosure of the international application.

Alternatively, the applicant may file a demand for International Preliminary Examination and submit to the International Preliminary Examining Authority (IPEA) a written submission in response to the written opinion, rebutting the objections raised by the examiner together with, where appropriate, amendments to the claims and specification.[21] The above demand must be filed within the above stipulated timeframe, together with the payment of a prescribed fee. The written opinion of the ISA is considered to be the written opinion of the IPEA where the IPEA acts as the ISA.[22] The IPEA may issue a second opinion based on the written submission by the applicant, and the applicant is given a second opportunity to rebut the examiner's objections by written submission, and where appropriate, amendments to the claims and specification. When a demand is filed, the IEPA issues an IPRP Chapter II of the PCT.[23]

However, if the above demand is not filed, an International Preliminary Report on Patentability (IPRP) Chapter I of the PCT is issued by the IB on behalf of the ISA.[24]

The IB communicates the IPRP to all elected offices.[25] Having a favorable IPRP no doubt increases the chances of obtaining a patent in the country where the applicant intends to seek patent protection during the national phase entries stage.

National Phase

Prior to expiration of the 30 or 31 months deadline from the priority date or the international filing date, whichever is earlier, the applicant must enter the national phase of the elected contracting states where the applicant wishes to seek protection of the invention. It should be noted that, in order to secure a national phase entry into the contracting states of the applicant's choice, the patent application must fulfill all the requirements, and the prerequisite official fee set out by the respective elected offices must be paid.

Once the applicant enters the national phase, the patent office of the designated states are responsible for granting the patent application. In Singapore, a national phase entry for an international application must be initiated 30 months from the priority date or the international application date, whichever is earlier.

Trademark Registration in Singapore

This section provides a brief description of the trademark registration process in Singapore. In summary, the trademark application process can be subcategorized into five major steps consisting of:

1. Filing an application
2. Formalities examination
3. Search and examination
4. Publication and opposition proceedings
5. Trademark registration

Filing an Application

The applicant may file an application for trademark registration by submitting the prescribed form, paying the prescribed fee[26] and by furnishing the following information to the Registry of Trade Marks, IPOS:

(a) A request for registration of a trademark
(b) Name and address of the applicant
(c) A clear representation of the trademark
(d) A list of goods or services in relation to which trademark registration is sought
(e) A declaration of use/intent to use the trademark

Formalities Examination

Upon receiving the filing of an application for a trademark, the Registry of Trade Marks proceeds to conduct a formalities examination to ensure that the trademark application forms and representations of the trademark submitted fulfill the requirements of the Trade Marks Act.[27]

If the trademark application is in order and has included the correct prescribed official fee, an application number and date is assigned by the Registry of Trade Marks.[28]

Search and Examination

Once the trademark application has been accorded an application number, a trademark search and examination ensues. In short, the trademark examiner conducts an examination of the trademark making sure that this said trademark satisfies the registration requirements as set out in the Trade Marks Act. A registrable trademark should possess the following:

(a) Distinctive–novel and comprises of apparent features
(b) Distinguishable–capable of being differentiated from the others which provide similar goods or services[29]

During this second stage, the registrar performs an indigenous search in the Singapore trademark database for conflicting marks. This is to ensure that no identical or similar mark has been registered or filed prior to the applicant's mark. This is an extremely crucial step, as the employment of similar or nearly similar marks may lead to confusion of the general public. If the trademark examiner discovers a registered trademark or a trademark pending registration which is similar to the applicant's mark, the latter may be refused registration by the trademark examiner. An office action is issued by the Registry of Trade Marks explaining the reasons for the refusal, and the applicant is given a specified period of time in which to put forth his argument in response to the said office action.[30]

Further, the registrar also checks if the specification of goods and services as sought by the applicant conforms with the International Classification of Goods and Services, and the Registry of Trade Marks then issues an office action if the said specification does not conform in the necessary manner. The applicant is given a stipulated period of time to file a response to the Registrar's refusal.[31] This response is evaluated by the registrar to ascertain if the reasons given or the arguments presented by the applicant are justifiable and valid enough to overcome the refusal. However, if an incorrect classification of goods and services is made, the Registry of Trade Marks rejects the application, and the applicant may need to re-file a new application.

Once the refusal is overcome, or if there is no objection raised by the trademark examiner, the trademark application proceeds to publication.

Publication and Opposition Proceedings

After the application for trademark registration is accepted, it is published in the Singapore Trade Marks Journal. The trademark is published for a span of 2 months, and throughout this time any interested members of the public, or a third party with a valid ground of opposition, may file a Notice of Opposition with the Registry of Trade Marks to protest the registration of the trademark.[32]

Upon receiving a Notice of Opposition, the applicant shall produce a counterstatement justifying his rights to the use of the trademark. Non-filing of a counterstatement results in the trademark being withdrawn. Upon receiving the counterstatement, the opponent files evidence in support of the opposition with the Registry of Trade Marks. Similarly, the applicant shall file evidence in support of the application upon receipt of the opponent's evidence with the Registry of Trade Marks. The opponent is given another opportunity to file evidence in response to the applicant's evidence in support of the application. The filing of all the evidence by the applicant and/or the opponent is by way of a statutory declaration. After all the evidence has been filed, the registrar gives notice to the applicant and the opponent to attend a hearing, wherein both parties present their respective cases and the evidence therein. If the outcome of this hearing is in favor of the applicant, the trademark proceeds to registration; otherwise, the trademark registration is refused.

On the other hand, if the trademark is published without contention from any third party, it proceeds to registration at the conclusion of the opposition period.

Trademark Registration

The final stage of the trademark application is the registration of the trademark. Upon expiry of the opposition period, and if there are no opposition proceedings, the trademark proceeds to registration and a certificate of registration issued. Unlike granted patents, the term of a registered trademark is 10 years[33] from the date of filing and the protection of the said trademark continues ad infinitum provided that the renewal fee is duly paid every 10 years and the trademark is in proper and continuous use.[34]

Filing International Registration of Trademark

Similarly to patent protection, trademark protection is also territorial in nature and applicants seeking trademark protection in various jurisdictions must submit multiple applications to the countries of their choice. The Protocol Relating to the Madrid Agreement Concerning the International

Registration of Marks (the Madrid Protocol), administered by the IB of the WIPO, enables an applicant to seek international trademark registration in various member states. In this respect, the Madrid Protocol provides a simple, one-stop solution for filing a trademark registration in multiple jurisdictions, wherein the applicant needs to deal with only one application to a single office, in one language, with one prescribed form, and one set of fees.[35]

An applicant who intends to file an international trademark registration via the Madrid Protocol is required first to file an initial trademark application for the same mark with the Intellectual Property Office (IPO) in the country of origin.[36] The country of origin in this case must also be a member state of the Madrid Protocol. In Singapore, an applicant must first file an application for the trademark with the Office of Origin, that is, IPOS.

As Singapore is a member state of the Madrid Protocol, the applicant may, within six months from the date of filing of the Singapore trademark application, submit an international trademark application to the IB through the Office of Origin (that is, IPOS). The submission shall be accompanied by the prescribed application forms designating the countries where protection is sought. IPOS will examine the application for irregularities upon receipt of the international application. To rectify any irregularities, IPOS informs the applicant. IPOS submits the international application to IB of the WIPO when the international application is in order.

The IB examines the international application to ensure that it conforms with the requirements of the Madrid Protocol. Once the trademark application is ascertained to be in compliance with all such requirements, an international registration number is issued. The filing date is the date on which IPOS receives the international application. The IB notifies IPOS and the designated member states of the international registration and publishes the particulars in the WIPO Gazette of International Marks.[37] A certificate of registration is also issued to the applicant. One advantage of filing an international trademark application through the Madrid Protocol is that all designated member states enjoy the same protection, as if the trademark application had been filed directly in the trademark offices of these countries.

Thereafter, the prosecution for the registration of the trademark proceeds as per the common process of a national trademark application in each of the designated member states until the point of registration. The trademark is subject to examination before the trademark office of the respective designated member states.

However, it should be made clear to the applicant that for a period of five years from the date of application for the trademark with the Madrid Protocol, the international trademark application is largely dependent on the initial trademark application that was first filed with the office of origin.[38] Having said that, abandonment, rejection, or the cancellation of the trademark originally filed with the office of origin renders the international trademark application null and void. After the expiration of the five-year period, the international trademark application becomes independent of the original trademark application first filed with the office of origin.

Summary

Patent is a limited monopoly right granted to a patent proprietor of an invention which is novel, comprises an inventive step, and has industrial applicability. Patent protection is territorial in nature, hence, a patent proprietor may only commence legal proceedings against a third party for infringing his patent in the country in which patent protection is obtained.

A complete patent application in Singapore includes the submission of the prescribed forms and fees, together with a copy of an English specification describing the invention. Upon receiving a complete application, the patents registry issues an application number and an application filing date. The subsequent dates for further prosecution of the application are calculated from the said application filing date.

As discussed above, patent protection is territorial in nature; hence, a patent application needs to be filed in every country in which the applicant wishes to seek protection. This is largely subject to the discretion of the applicant wherein a commercial decision will need to be made. Alternatively, an applicant may opt to file an *international* application with the Patent Cooperation Treaty (PCT), a Treaty governed by the World Intellectual Property Office. Singapore is a contracting state of the PCT and the Intellectual Property Office of Singapore (IPOS) acts as the receiving office for all *international* applications filed by a resident of the country.

It should be noted that prior to filing a first patent application in a foreign country, no person resident in Singapore shall, without written authority granted by the patent registrar, file or cause to be filed outside Singapore an application for a patent of an invention.[39]

When an applicant first files a patent application in Singapore, he may proceed to file the patent application overseas for the same invention if there is no direction issue under Section 33 of the Patents Act after two months from the date of filing of the Singapore patent application. In this case, he may assume the patent registrar has reviewed the patent application and that there is no security concern.

A granted patent has a maximum lifespan of 20 years from the application filing date provided that the yearly renewals are continued. After the expiration of 20 years, the patent falls into the public domain wherein it is opened to manipulation and use by the general public.

In Singapore, the registration of trademarks is not mandatory. Protection of a trademark is largely subject to the discretion of the business owner. A registrable mark should be distinctive and distinguishable. In other words, it should be novel, comprised of perceptible features, and should not resemble other marks that provide similar goods and services.

To protect a mark, an application must be filed with the Trade Mark Registry. The mark is subjected to preliminary examination to ascertain that all requirements are met. An application number and filing date is issued once the registrar confirms that the said application fulfills the requirements of the Registry. Subsequently, the mark is researched and examined to ensure that it satisfies the registration requirements set by the Trade Mark Acts and that no similar marks have been registered or filed earlier.

Prior to acceptance of the mark by the Trade Mark Registry, it is published for public inspection and opposition for a period of two months. In absence of oppositions, a Certificate of Registration is issued to the applicant. Once a mark is registered, it can be maintained indefinitely provided a renewal fee is paid every 10 years.

A mark may be protected by filing an international application through the Madrid Protocol administered by the International Bureau of the World Intellectual Property Office. However, an applicant must first file an application in the country of origin which shall be a member state of the Protocol. Singapore is a member state of the Madrid Protocol and the Intellectual Property Office of Singapore (IPOS) acts as the office of origin, wherein all applicants intending to file an international application shall first file an indigenous application in Singapore. Within six months from the date of filing in the office of origin, the applicant may then proceed to file an international application under the Madrid Protocol. The initial five years of the international application are dependent on the status of the mark as originally filed with the office of origin. Therefore, rejection of the mark originally filed in the office of origin leads to the revocation of the international application.

Seeking protection for patents and trademarks is nonobligatory. However, a well-protected invention and mark provides the applicant with an exclusive right to control the use, reproduction, distribution, and further manipulation of the invention or mark in various forms.

8

Case Reviews

Introduction

This chapter provides case reviews of several important intellectual property lawsuits in Singapore. These cases are selected for discussion because they provide a good illustration of the various legal principles enunciated in the earlier chapters of this book. Some of these legal principles of patent, trademark, and copyright law can be difficult to understand and, for this reason, they are best explained through cases so that one can better appreciate and understand how these legal principles are being applied in legal suits.

Case Review One: Trek Technology (Singapore) Pte Ltd v. FE Global Electronics Pte Ltd and Others and Other Suits (No. 2) (2005) SGHC 90

Legal Principles

This case illustrates the principles of ownership and assignment in relation to patents, in particular:

(a) The right to be granted a patent under the Patents Act
(b) The requirements for a valid assignment under the Patents Act

Material Facts and Background of Action

The plaintiff (*Trek*) in this case commenced actions against several defendants (*M-Systems, FE Global, Electec,* and *Ritronics*) for various infringements of its patent for a data storage device known as the ThumbDrive (the *Patent*).

Figure 8-1

In their defense, the defendants argued among other things, that Trek had made misrepresentations concerning the inventorship and/or ownership of the Patent and thus sought to revoke the Patent on that ground. M-Systems, in particular, claimed that:

(a) The ThumbDrive was invented by one Marcus Cheng (*Mr. Cheng*) and one Poo Teng Pin (*Mr. Poo*), who were employees of Trek and its related company (*S-Com*). Trek was thus not the full owner of the Patent.
(b) There was a misrepresentation to the Registrar of Patents when Trek did not name Mr. Poo as one of the inventors when it applied for the Patent.

High Court's Decision in Relation to the Above Facts

The Court noted that to revoke a patent on the ground of misrepresentation, the misrepresentation must be material. This principle was enunciated in the cases of *Intalite International N.V.* v. *Cellular Ceilings Ltd (No. 2) (1987)] RPC 532 and Speedy Gantry Hire Pty Ltd* v. *Preston Erection Pte Ltd (1998) 40 IPR 543*. The onus was, thus, on the defendants to show that the misrepresentation had actually deceived the registrar of patents into granting the Patent.

The Court found that the inventive concept of the ThumbDrive was developed by Mr. Cheng and Mr. Poo, two employees of S-Com and Trek. S-Com and Trek were wholly owned subsidiaries of the parent company, Trek 2000 International. The Court found that there was some evidence of an agreement or understanding between both companies that the patents were to be registered in the name of Trek.

From the facts, the Court did not find evidence to suggest a deliberate omission by Trek or Mr. Cheng of Mr. Poo's name from the patent application. The cause of the omission, as the Court found, was *inadvertence.* The Court was also satisfied that once Trek discovered the omission of Mr. Poo's name as a co-inventor in the patent application, Trek "acted with reasonable dispatch" to rectify the error by notifying the World Intellectual Property Organization (WIPO) and the Intellectual Property Office of Singapore (IPOS).

Further, the Court found that upon discovery of the mistake, Trek entered into an assignment agreement with S-Com for S-Com to transfer whatever residual rights it had in the invention to Trek.

At this juncture, the Court highlighted the contrast between categories of parties who are entitled to apply for the Patent (under Section 19(1) of the Patents Act), and parties who are entitled to the grant of the Patent (under Section 19(2) of the Patents Act).

Under Section 19(2) of the Patents Act:

"A patent for an invention may be granted:

(a) Primarily to the inventor or joint inventors
(b) In preference to paragraph (a), to any person or persons who, by virtue of . . . an enforceable term of any agreement entered into with the inventor before the making of the invention, was or were at the time of the making of the invention entitled to the whole of the property in it (other than equitable interests) in Singapore
(c) In any event, to the successor or successors in title of any person or persons mentioned in (a) or (b) or any person so mentioned and the successor or successors in title of another person so mentioned

and to no other person."

The Court also distinguished between assignment of *any right in patent* and assignment of the rights to an invention. In the latter, which is the category that the present case falls within, the assignment agreement does not need to be formally executed. On the other hand, in the case of assignment of *any right in patent,* formal execution (in writing, signed by both parties) is required under Sections 41 and 43 of the Patents Act.

Applying the above provisions to the facts of the case, the Court found that Trek's right to be granted the Patent was derived from two sources:

(a) Mr. Cheng, its employee, pursuant to Section 19(2)(b) of the Patents Act
(b) The assignment by S-Com of its right to the invention in favor of Trek. This would entitle Trek, pursuant to Section 19(2)(c) of the Patents Act, to be granted the Patent as a successor-in-title to S-Com.

The Court further addressed the question of whether the factual inaccuracies in Trek's application process materially contributed to the decision to grant the Patent. The Court answered in the negative by referring to the decision of the English Patents Court in *Coflexip Stena Offshore Limited's patent(1997) RPC 179* where it was pronounced that, "What really matters is that the register should show who the proprietor is. How he came to be the proprietor is of no or little importance."

Conclusion and Commentary

It becomes apparent that M-Systems' claims (as outlined previously) to have the Patent revoked in this case failed because:

(a) The Court held that the evidence was sufficient to show that Trek had valid ground to hold itself out to the Registrar of Patents as the owner of the Patent. The assignment agreement in particular carried *evidentiary weight* in that it clarified "beyond doubt that S-Com had transferred all rights in the invention to Trek." Trek was thus entitled to file the patent application as the owner.

(b) The Court did not find that there was a misrepresentation by Trek, as the omission of Mr. Poo, and Trek's subsequent rectification of that mistake could not be said to materially affect the grant of the Patent.

Court of Appeal's Decision

The defendants appealed against the decision (reported in *FE Global Electronics Pte Ltd and others* v. *Trek Technology (Singapore) Pte Ltd and another appeal (2006) 1 SLR 874*) on other issues related to the patentability of the Patent (ownership and assignment of the Patent were not issues raised during the appeal). The Court of Appeal upheld the High Court's decision and dismissed the appeal.

Case Review Two: Creative Technology Ltd v. Aztech Systems Pte Ltd (1997) 1 SLR 621

Legal Principles

The following case illustrates the concept of ownership of copyright and is particularly useful in explaining how the ownership in copyright differs from that of ownership in a patent.

Material Facts and Background of Action

The case was an appeal from the judgment of the High Court (*Aztech Systems Pte Ltd* v. *Creative Technology Ltd [1996] 1 SLR 683*). Creative Technology (*Creative*), the defendants in that action, alleged that Aztech Systems Ptez (*Aztech*) infringed its copyright in various firmware and software by copying Creative's programs. The salient facts are:

(a) Sound cards. The dispute centered on sound cards. A sound card is essentially a circuit board that can be inserted into a personal computer (PC). The sound card generates sounds through a speaker system by means of a digital signal processor (DSP) and a digital-to-analog converter (DAC). When an application program runs on a PC and requires sound, it sends commands to a software program in a driver for conversion into low-level commands. These commands are sent to the DSP and, subsequently, on to the DAC where the data is converted into sound through the speaker system.

(b) Creative sound cards. The subject matter of this action concerned sound cards developed by Creative (*Sound Blaster*) which used the Intel 8051 chip. Creative wrote a computer program (*firmware*) for this chip (*Creative chip*). The firmware program was burnt into the read-only memory (ROM) of the Creative chip. Creative also bundled its Sound Blaster with other software, including TEST-SBC and the driver called CT-VOICE.DRV.

(c) Aztech sound cards. Aztech were developers of sound cards from the *Sound Galaxy* family (in three versions, *Sound Galaxy BX*, *Sound Galaxy NX* and *Sound Galaxy NX PRO*), which were compatible or interoperable with the application programs that had been developed to operate with Sound Blaster.

(d) Creative alleged that for the development of Sound Galaxy BX, Sound Galaxy NX, and Sound Galaxy NX PRO, Aztech infringed its copyright in the programs in the firmware in Creative's chip, as well as CT-VOICE.DRV and TEST-SBC by copying the whole or parts of Creative's programs.

(e) Aztech denied infringement on the ground that, inter alia, it had a right to use the Sound Blaster sound card and the associated software TEST-SBC and driver CT-VOICE.DRV under the principle enunciated in the case of *Betts* v. *Willmott (1871) 6 LR Ch App 239* (the *Betts principle*). This principle states that a purchaser of an article should be expected to have control of it or to use it as he pleases. Aztech argued that its purchase of a Sound Blaster sound card, together with the TEST-SBC and CT-VOICE.DRV, made it the lawful owner of the goods with the presumed and implied right of quiet enjoyment and possession. This included the right to use them for any reasonable purpose and to use them with a DEBUG program to study its mode of operation in order to understand the various commands, with a view to designing a noninfringing Sound Blaster compatible sound card.

High Court's Decision in Relation to the Above Facts

The High Court ruled in favor of Aztech and accepted the ground it put forth. It was held that Aztech, as it had purchased a Sound Blaster sound card together with TEST-SBC, was entitled in the exercise of its rights of ownership over it to use it by running it with the program DEBUG to study its mode of operation, and with a view to designing a Sound Blaster compatible sound card. The Court further held that the entitlement to do so stemmed, not from the fact that such use was reasonable, but because such use was a right of ownership conferred on Aztech by virtue of its purchase and there was nothing in the facts to suggest that Creative had no power to confer this right.

Court of Appeal's Decision

Creative appealed against the High Court's finding that Aztech had a right of ownership over Creative's software programs by virtue of its purchase of the Sound Blaster sound card.

The Court of Appeal, in allowing the appeal, reversed the High Court's ruling and disagreed with the High Court's acceptance of the Betts principle. The case was distinguished on the ground that it was a patent case, and the proposition contained therein cannot be extended to copyright. This is because exclusive rights granted to the patent owner can differ materially from those granted to the copyright owner. Exclusive rights which are granted to the holder of the copyright do not include rights to use and sell the protected work.

The Court of Appeal highlighted the difference by referring to Section 66(1) of the Patents Act and Section 26(a) of the Copyright Act. Under Section 66(1) of the Patents Act the exclusive rights granted to the patentee include *use* or *disposal*. By contrast, Section 26(a) of the Copyright Act grants to the copyright owner of a literary work the exclusive rights to reproduce, publish, perform, broadcast, cable-cast, make an adaptation, and to do all the said acts in relation to an adaptation of the work. The Court of Appeal held that in copyright law, rights only exist insofar as they are provided for by the Copyright Act. The Copyright Act does not, for instance, permit the purchaser of a computer program to use it to multiply and sell infringing copies, since these are infringing acts under the Copyright Act unless performed with the permission of the copyright owner.

The Court of Appeal also agreed with Creative's counsel that when a patented article is sold to a purchaser, the patentee has no legitimate expectation or interest in controlling the further use of the article by the purchaser: "effectively there is nothing left of the patentee's rights." On the other hand, a copyright owner who sells his or her work still retains the economic and moral rights of authorship. Therefore, the Court of Appeal concluded that the proposition in *Betts* v. *Willmott* cannot be applied in the context of copyright law in Singapore. The purchase of the Sound Blaster sound card did not imply consent by Creative to the copying of TEST.SBC by Aztech, in order to discover its functionality and with the objective of making a competing product. The Court of Appeal's opinion was that to uphold such an implied license would be tantamount to "making a mockery" of the provisions of the Copyright Act.

Case Review Three: Real Electronics Industries Singapore (Pte) Ltd v. Nimrod Engineering Pte Ltd (T Vimalanathan, Third Party) (1996) 1 SLR 336

Legal Principles

The following case demonstrates the issue of ownership and assignment in copyright cases. In particular, the case makes it clear that only the owner or legal assignor of copyright has the right to bring an infringement action.

Material Facts and Background of Action

The plaintiff (*REIS*) and defendant (*Nimrod*) in this case were fax modem manufacturers. REIS claimed that a fax modem marketed by Nimrod infringed its copyright in the printed circuit board design of the *COMFAX CF-29SR* fax modem marketed by REIS. According to REIS, the printed circuit board layout and the artwork design of their modem were the product of their consultant engineer, one Mr. Soh.

In their defense Nimrod argued, inter alia, that REIS did not have the right to sue as they did not have ownership or title to the design. The copyright in the design was vested in Mr. Soh, but Mr. Soh was not an employee of REIS. On this ground, Nimrod relied on Section 194(3) of the Copyright Act which states that no assignment of copyright would have effect unless it was in writing signed by or on behalf of the assignor.

High Court's Decision in Relation to the Above Facts

The Court found in favor of REIS on the issue of ownership. From the facts, the Court found that Mr. Soh had explicitly conceded that the design he developed was the property of REIS. Thus, as the Court concluded, "given the scenario, the issue of assignment, as envisaged under Section 194(3) of the Copyright Act, does not arise at all in the case at hand"

Conclusion and Commentary

The Court's brief discussion on the issue of ownership and assignment in this case was a result of the peculiar set of evidence and facts that clearly indicated that ownership resided with REIS. However, the Court acknowledged that the point of law on ownership and assignment was of *importance and significance.* Reference was made to The Law of Copyright in Singapore by Professor George Wei and the cited case of *Performing Right Society Ltd* v. *London Theatre of Varieties (1924) AC 1* where, among other things, it was stated that, "until the execution of a legal assignment occurs, the equitable assignee would not be entitled to maintain an action for copyright infringement in his own name and that he would have to join the legal owner as the joint plaintiff." Thus, what the Court was, in effect, suggesting was that any copyright infringement action may be challenged if ownership or proper assignment of the copyright in question fails to be established by a plaintiff.

Case Review Four: Seiko Epson Corp v. Sepoms Technology Pte Ltd and Another (2007) 3 SLR 225

Legal Principles

The following case illustrates the working of Section 69(1) of the Patents Act that states that the period of liability for damages and/or an account of profits in a patent infringement action is restricted to the period when the infringer knew or had reason to suppose that an infringed patent existed.

Material Facts and Background of Action

The plaintiff manufactured ink jet printers. The defendants manufactured and sold compatible and refillable ink cartridges, which could be used to replace ink cartridges such as those manufactured by the plaintiff. The plaintiff and the defendants both owned patents relating to ink cartridges.

The plaintiff commenced a suit against the defendants on October 5, 2005, for patent infringement of Singapore Patent No. SG46602 (the *Pat-*

ent). On August 2, 2006, the parties reached a consent judgment in which the defendants admitted liability for the infringement of the Patent. The terms of the consent judgment included an account of profits by the defendants.

In compliance with that term, the defendants filed in Court their accounts commencing from the date of the writ (October 5, 2005) to July 31, 2006 (the *accounted period*). The defendants' director deposed in his affidavit of evidence-in-chief that the defendants first knew of the Patent on October 7, 2005, when they were served with the statement of claim in the suit. Prior to that, the defendants did not receive any *cease and desist* letter from the plaintiff, nor were they informed orally or in writing by the plaintiff of the existence of the Patent or of the defendants' infringement. Thus, the defendants' director was of the view that the defendants were only liable to account to the plaintiff from October 7, 2005, to August 2, 2006.

The plaintiff was dissatisfied with the defendants' accounts as filed, and filed an objection against the accounts, pointing out that the accounted period differed from the period of infringement, which according to the plaintiff was from the date of publication of the Patent (February 20, 1998) until the date of the consent judgment. The plaintiff also objected to the affidavit of the defendants' director, more specifically, he had an issue with those paragraphs in which the director had related the events from which the defendants acquired the requisite knowledge of the Patent.

The plaintiff followed up on its objections by filing a summons (the *Application*) requiring the defendants to file a further set of accounts for the period February 20, 1998, to September 30, 2005. The assistant registrar dismissed the Application with costs to the defendants, deciding that a summons hearing was not an appropriate forum for a determination of the date when the account of profits by the defendants should commence. The determination was best left to the Registrar conducting the inquiry as to the appropriate amount of damages to be paid by the defendants to the plaintiff under the consent judgment.

The plaintiff in the present action appealed against the Assistant Registrar's decision.

High Court's Decision in Relation to the Above Facts

The High Court dismissed the plaintiff's appeal on three grounds. First, under Section 69(1) of the Patents Act, the defendants' liability for damages for infringement and/or an account of profits was limited to the period when they knew or had reasonable grounds for supposing that the infringed patent existed. The defendants, therefore, had the right to assert and, later at the inquiry prove that they had no knowledge of the Patent's existence until they were served with the writ in the suit.

Secondly, the High Court held that the consent judgment was final only on the issue of liability for patent infringement by the defendants. The consent judgment was akin to an interlocutory judgment, leaving damages to be assessed, because the issue of profits due to the plaintiff resulting from the infringing acts of the defendants had to be resolved later at an inquiry to be conducted by the Registrar. Just like a hearing before the Registrar for assessment of damages pursuant to an interlocutory judgment, evidence would have to be adduced at the inquiry, and witnesses would have to testify on the issue of when the defendants acquired knowledge of the Patent. It was therefore premature of the plaintiff to raise objections at this stage on the accounts filed by the defendants.

Thirdly, the Court also considered the *unduly onerous and perhaps unnecessary burden* upon the defendants if the Application was allowed at that stage. The costs of gathering old accounting records should only be incurred after the defendants had been found liable, at the inquiry, to account for the entire period February 20, 1998 to August 2, 2006. Otherwise, such costs would have been needlessly incurred by the defendants if the Registrar ruled at the inquiry that the defendants were not required to account for any period before October 7, 2005. Furthermore, any prejudice to the plaintiff that may arise from disallowing the Application before the inquiry could be easily cured by compensation through an appropriate order of costs against the defendants.

Thus, the High Court held that granting the Application was premature before the inquiry stage and dismissed the plaintiff's appeal with costs.

Conclusion and Commentary

The plaintiffs appealed against the decision (reported in *Seiko Epson Corporation* v. *Sepoms Technology Pte Ltd and another (2008) 1 SLR 269.*) arguing that the effect of the consent judgment (and the order stating that there would be an account of profits by the defendants) was that it was implicit that the defendants had agreed to having infringed the Patent from the date alleged by the plaintiff in its claim. However, the Court of Appeal rejected this argument, upholding the High Court decision that an objective construction of the consent judgment indicated that it was final only on the issue of liability, with the specific accounting period to be determined at a later stage.

Case Review Five: Wing Joo Loong Ginseng Hong (Singapore) Co Pte Ltd v. Qinghai Xinyuan Foreign Trade Co Ltd and Another (2008) 3 SLR 296

Legal Principles

The following case illustrates the issues that may arise with the assignment of trademarks. Section 38(1) of the Trade Marks Act provides that, "a registered trademark is assignable and transmissible in the same way as other personal or movable property, and is so assignable or transmissible either in connection with the goodwill of a business or independently."

Registration of an assignment is not mandatory under the Trade Marks Act but nonregistration may lead to two consequences. First, according to Section 39(3) of the Trade Marks Act, until an application has been made for the registration of the assignment, the assignment is ineffective as against a person acquiring a conflicting interest in or, under the registered trademark, in ignorance of the transaction.

For the purposes of illustration, if Roger assigns a trademark to Sunny and subsequently Roger also grants an exclusive license to Timothy, Sunny would take the assignment subject to Timothy's license if Sunny does not register his assignment.

The second consequence of nonregistration as provided under Section 39(4) of the Trade Marks Act is that the assignee, if he brings an action for infringement, is not entitled to claim for damages an account of profits or statutory damages in respect of that infringement.

Further, it is noteworthy that applications to register the assignment of a trademark (to prevent either of the two undesirable consequences above from happening) may be challenged, if fraud or misrepresentation were involved.

Material Facts and Background of Action

In this case, the *Rooster* trademark was registered in Singapore with effect from September 6, 1995, by a certificate issued on August 27, 2001. The first registered owner of the trademark was Qinghai Medicines & Health Products Import & Export Corp. of Qinghai, China (*Qinghai Meheco*).

The trademark went through two assignments: (a) to the first defendant and another entity, Qinghai Yixin Medical Co. (*Qinghai Yixin*); and (b) subsequently, Qinghai Yixin assigned its rights to the trademark to the first defendant, thereby resulting in the first defendant becoming the sole owner of the trademark.

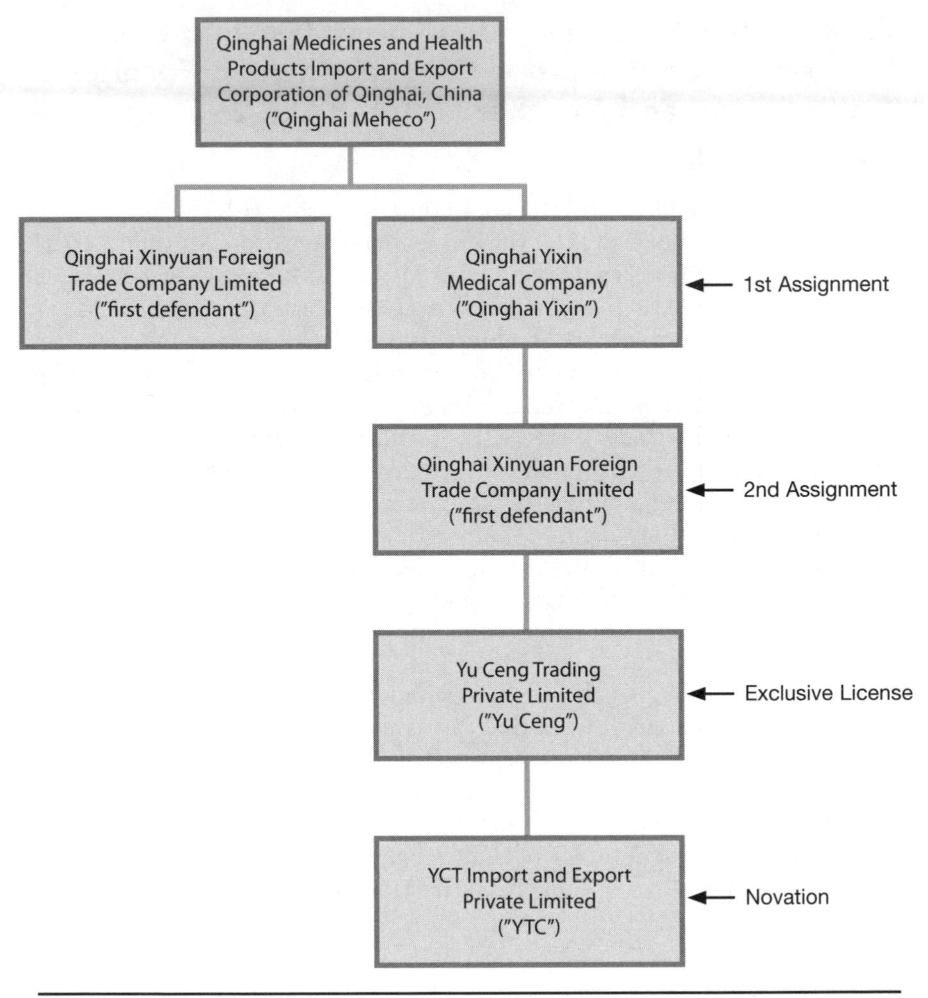

Figure 8-2

The first defendant applied to the Singapore Registry of Trade Marks to transfer the Singapore trademark to its name, and the transfer was recorded in 2006, backdated to 2003. In 2005, the first defendant granted an exclusive license to Yu Ceng Trading Pte (*Yu Ceng*) to use the Singapore trademark.

In 2005, Yu Ceng applied for and obtained search warrants pursuant to which the plaintiff's premises were raided, and quantities of cordyceps with alleged counterfeit *Rooster* trademarks were seized. In 2006, Yu Ceng's license was novated in favor of the second defendant, YCT Import & Export Pte (*YCT*).

In the present proceedings the plaintiff sought, among other things, an order that the trademark registration be revoked or declared invalid and a declaration that any copyright in the *Rooster* labels or in any literary or artistic work in each of the labels did not subsist in favor of the defendants, and that the plaintiff had not infringed any copyright. One of the arguments put forth by the plaintiff to invalidate the trademark was that the first defendant had obtained the registration of the *Rooster* mark in its name in 2005 and 2006 through fraud and/or misrepresentation to the registrar of trademarks.

High Court's Decision in Relation to the Above Facts

The Court found that the thrust and focus of the plaintiff's complaint was that registration of the two assignments was marred by fraud and misrepresentation. However, the plaintiff's applications were to have the registration of the trademark revoked and invalidated under Section 23(4) of the Trade Marks Act. The complaint (on fraud and misrepresentation concerning the assignments) was thus outside the scope of the proceedings. Despite this, the Court went on to state its views on the matter.

The Court noted that during the hearing, the plaintiff and the defendants had called lawyers from China to give evidence on the state, effect, and operation of Chinese corporate and insolvency law on whether, among other things, the trademark rights were in fact assigned with the other assets of Qinghai Meheco pursuant to a joint acquisition agreement between Qinghai Meheco, the first defendant, and Qinghai Yixin. After examining the evidence and the submissions of counsel, the Court found itself unable to come to a clear conclusion whether the trademark rights were assigned, largely because the uncertainties raised by the plaintiff were essentially matters which had to be determined under Chinese law and procedure.

The Court thus held that the issue was best resolved by seeking a formal decision from a competent court in China, but that on the evidence, the Court was satisfied that the plaintiff had not proved that the applications to register the assignments were affected by fraud or misrepresentation.

Conclusion and Commentary

The plaintiff appealed against the decision of the High Court (reported in *Wing Joo Loong Ginseng Hong (Singapore) Co. Pte Ltd* v. *Qinghai Xinyuan Foreign Trade Co Ltd and another and another appeal (2009) 2 SLR 814)*. The Court of Appeal, however, came to the same conclusion as the High Court and held that the complaint of fraud was outside the scope of proceedings. This was because Section 23(4) of the Trade Marks Act only applied to the *original* registration of a sign or mark as a trademark, and not the registration of *subsequent* transactions such as assignments of a registered trademark.

Even if it was within the scope of proceedings, the Court of Appeal further held that to succeed in invalidating the registration of the trademark on grounds of fraud or misrepresentation, it must be shown by the plaintiff that the registration by the defendants succeeded only on the strength of an untrue statement, which the plaintiff had failed to do in this case.

Thus, although the decision of the Courts in this case was determined largely by the fact that the plaintiff's applications were to have the registration of the trademark revoked and invalidated (and not the registration of the assignments), it was acknowledged that a finding of fraud or misrepresentation would constitute a legitimate challenge to the validity of registration of the assignment of the trademark.

Summary

This chapter examined five cases on intellectual property decided by the local courts that illustrate principles discussed in previous chapters. Case 1 (*Trek Technology (Singapore) Pte Ltd* v. *FE Global Electronics Pte Ltd & Other Suits (No. 2) [2005] SGHC 90*) discusses the right to the grant of a patent, and the requirements of a valid assignment of a patent. In that case the Court also dealt with fraud and misrepresentation and the requirements of a successful claim in relation to the same.

The same issues of fraud, misrepresentation, and assignment were also dealt with by the Court in Case 5 (*Wing Joo Loong Ginseng Hong (Singapore) Co Pte Ltd* v. *Qinghai Xinyuan Foreign Trade Co Ltd and Another [2008] 3 SLR 296*), but this time pertaining to trademarks. What may be gleaned from this case is that registration of an assignment of a trademark is certainly recommended to avoid confusion and disputes in future dealings.

While Cases 1 and 5 dealt with patents and trademarks respectively, Case 2 (*Creative Technology Ltd* v. *Aztech Systems Pte Ltd [1977] 1 SLR 625*) touched on copyright issues, particularly ownership of copyright, and how it differs from ownership in a patent.

Case 3 (*Real Electronics Industries Singapore (Pte) Ltd* v. *Nimrod Engineering Pte Ltd*) continued on the issues related to copyright, as the Court discussed principles of ownership and assignment in a copyright case. The case makes it clear, for instance, that only the owner or legal assignor of copyright has the right to bring an action for infringement.

On matters pertaining to remedies, Case 4 (*Seiko Epson Corp* v. *Sepoms Technology Pte Ltd and Another [2007] 3 SLR 225*) highlighted that the period of liability for damages and/or an account of profits in a patent infringement action is restricted to the period when the infringer knew or had reason to suppose that an infringed patent existed.

Endnotes

Chapter 1

1. Bardeen, J. and W. Brattain. "Three-Electrode Circuit Element Utilizing Semiconductive Materials." U.S. Patent 2,524,035. October 3, 1950.
2. Shockley, W. "Circuit Element Utilizing Semiconductor Material." U.S. Patent 2569347. September 25, 1951.
3. Bardeen, J. "Semiconductor Research Leading to the Point Contact Transistor." Nobel Lecture. December 11, 1956. [Online] http://nobelprize.org/physics/laureates/1956/bardeen-lecture.pdf
4. IC Knowledge-History. [Online] http://www.icknowledge.com/history/1950s.html
5. Haviland, D. B. "The Transistor in a Century of Electronics." Nobel Prize. [Online] http://nobelprize.org/physics/educational/transistor/history/
6. Lilienfeld, J. E. "Method and Apparatus for Controlling Electric Currents." U.S. Patent 1745175. January 18, 1930.
7. Lilienfeld, J. E. "Device for Controlling Electric Current." U.S. Patent 1900018. March 7, 1933.
8. Heil, O. "Improvements In or Relating To Electrical Amplifiers and Other Control Arrangements and Devices." G. B. Patent 439,457. December 6, 1935.
9. Kilby, J. S. "Turning Potential into Realities." Nobel Lecture. December 8, 2000. [Online] http://nobelprize.org/physics/laureates/2000/kilby-lecture.html
10. "The Chip that Jack Built." Texas Instruments. [Online] http://www.ti.com/corp/docs/kilbyctr/jackbuilt.shtml
11. "The Hapless Tale of Geoffrey Dummer." *Electronic Product News*. October 1, 2005. [Online] http://www.epn-online.com/page/22909/the-hapless-tale-of-geoffrey-dummer-this-is-the-sad-.html
12. "Death of a Man Whose Idea Went on to Change the World Forever. *Worcester News*. September 17, 2002. [Online] http://archive.thisisworcestershire.co.uk/2002/09/17/250160.html

13. Noyce, R. "Semiconductor Device-and-Lead Structure." U.S. Patent 2,981,877. April 25, 1961.
14. Moore, G. E. "Cramming More Components onto Integrated Circuits." *Proceedings of the IEEE*, vol. 86, issue 1, (January 1998) pp. 82–85.
15. Annual Report. (2005). "2020 is Closer Than You Think." Semiconductor Industry Association. [Online] http://www.sia-online.org/galleries/annual_report/Annual%20Report%202005.pdf
16. Su, Y. H., R.-S. Guo, and W.-Y. Hsieh. "Historically Explore on the Patterns of Business Model for Silicon Intellectual Property (IP) Providers in the Semiconductor Industry." *Conf. Management of Engineering and Technology* (2005) pp. 447–458.
17. Dibiaggio, L. "Design Complexity, Vertical Disintegration and Knowledge Organization in the Semiconductor Industry." *Industrial and Corporate Change*, vol. 16, (2007) pp. 239–267.
18. Macher, J. T., D. C. Mowery, and T. S. Simcoe. "E-Business and Disintegration of the Semiconductor Industry Value Chain." *Industry and Innovation*, vol. 9, no. 3, (December 2002) pp. 155–181.
19. *2005 IC Economics Report*. Chapter 2: "IC Market Status and Trends." [Online] http://www.icknowledge.com/our_products/2005ICEconomics2.pdf
20. Greenagel, J. and A. Craib. "SIA Forecast: Global Chip Sales Will Surpass $321 Billion in 2010." Semiconductor Industry Association. November 14, 2007. [Online] http://www.sia-online.org/pre_release.cfm?ID=455
21. Greenagel, J. and A. Craib. "Global Chip Sales Hit Record $227.5 Billion in 2005." Semiconductor Industry Association. February 2, 2006. [Online] http://www.sia-online.org/cs/papers_publications/press_release_detail?pressrelease.id=36
22. Greenagel, J. and A. Craib. "Global Chip Sales Hit Record $247.7 Billion in 2006." February 2, 2007. [Online] http://www.sia-online.org/cs/papers_publications/press_release_detail?pressrelease.id=128
23. Buss, D., B. L. Evans, J. Bellay, W. Krenik, B. Haroun, D. Leipold, K. Maggio, Y. Jau-Yuann, and T. Moise. "SoC CMOS Technology for Personal Internet Products." *IEEE Trans. Electron Devices*, vol. 50, no. 3, (March 2003) pp. 546–556.
24. Buss, D. "Technology in the Internet Age." in *IEEE Int. Solid-State Circuits Conf. Dig. Tech. Papers*, (February 2002) pp. 18–21.
25. Dubai Silicon Oasis. [Online] http://www.dso.ae/

26. Wang, D. N. K. "What's Next for China's Emerging IC Industry." *Solid State Technology.* August 2005. [Online] http://sst.pennnet.com/Articles/Article_Display.cfm?Section=HOME&ARTICLE_ID=233889&VERSION_NUM=2&p=5
27. "IC Design Houses in China, 2005." *Market Research.* (September 16, 2005). [Online] http://www.marketresearch.com/product/display.asp?productid=1187420&g=1
28. Matthews, M. and T. Giovanetti. "Why Intellectual Property is Important." *IPI Ideas.* August 28, 2002. [Online] http://www.ipi.org/ipi%5CIPIPublications.nsf/PublicationLookupFullTextPDF/94061686270E14F286256C3800514943/$File/II-CaseForIP-2.pdf?OpenElement
29. Bugbee, B. W. "Genesis of American Patent and Copyright Law." Washington, DC: Public Affairs Press (1967).
30. Dobyns, K. W. "The Patent Office Pony: A History of the Early Patent Office." Sergeant Kirkland's Press (1994).
31. Shirer, M. and A. Havens. "Worldwide Software Piracy Rate Holds Steady at 35%, Global Losses Up 15%." Business Wire. May 15, 2007. [Online] http://findarticles.com/p/articles/mi_mOEIN/is_2007_May_15/ai_n27237837/?tag=content:coll
32. Phukan, S. and G. Dhillon. "Ethical and Intellectual Property Concerns in a Multicultural Global Economy." *Electronic J. Information Systems in Developing Countries,* vol. 7, no. 3, (2001) pp. 1–8.
33. Pele, A. F. "IP Business Model to Continue to Exist, Says Analyst." *EE Times.* June 12, 2007. [Online] http://www.eetimes.eu/uk/204702089
34. Heidarson, C. and J. Tully. "Forecast: Semiconductor Intellectual Property, Worldwide, 2005–2010." Gartner Dataquest. July 24, 2006. [Online] http://www.gartner.com/DisplayDocument?id=494233
35. Clarke, P. "Chip IP Market Set to Rise 25% in 2006, Says Gartner." *EE Times.* August 14, 2006. [Online] http://www.eetimes.com/showArticle.jhtml?articleID=191902209
36. Joselyn, L. "Picking the IP Winners." *Scottish Technology News (STN)—A Scottish Enterprise Publication,* issue 7, (August 2004).

Chapter 2

1. Machlup, F. "The Production and Distribution of Knowledge in the United States." Princeton, NJ: Princeton University Press (1962).

2. Organization for Economic Co-operation and Development (OECD). "The Knowledge-Based Economy." (1996). [Online] http://www.oecd.org/dataoecd/51/8/1913021.pdf
3. Hebner, R. "The Role of Photonics and Other Enabling Technologies in Driving Future Economic Growth." National Institute of Standards and Technology. October 23, 1997. [Online] http://www.nist.gov/speeches/optobos.htm
4. Scott, S. K. and G. Smalley. "Richest Man Who Ever Lived: King Solomon's Secrets to Success, Wealth and Happiness." Doubleday Publishing (2006).
5. Golden Gate Bridge—Construction data. [Online] http://goldengatebridge.org/research/ConstructionPrimeContr.php
6. Scott, M. D. "Scott on Information Technology Law." United States: Wolters Kluwer, Aspen Publishers (2006).
7. Brooktree Corporation vs. Advanced Micro Devices, Inc., 977 F.2d 1555 (U.S. Fed. Cir. 1992).
8. Rockman, H. B. "Intellectual Property Law for Engineers and Scientists." IEEE Press–Wiley Interscience (2004).
9. Small and Medium-sized Enterprises Division, Intellectual Property for Business, World Intellectual Property Organization (WIPO). http://www.wipo.int/sme
10. Ehmke, C. and J. Williams. "Intellectual Property: Obtaining Patents, Trademarks, and Copyrights." Purdue Extension-Agricultural Innovation and Commercialization Center. [Online] http://www.ces.purdue.edu/extmedia/EC/EC-723.pdf
11. Gajski, D. D., A. C.-H. Wu, V. Chaiyakul, S. Mori, T. Nukiyama, and P. Bricaud. "Essential Issues for IP Reuse." *Asia and South Pacific Design Automation Conf. (ASP-DAC)* (2000) pp. 37–42.
12. Chang, S. H. and S. D. Kim. "Reuse-Based Methodology in Developing System-on-Chip (SoC)." *Int. Conf. Software Engineering Research, Management and Applications* (August 2006) pp. 125–131.
13. Technology Working Groups. (2006, 2007). International technology roadmap for semiconductors 2006 update-Design. Semiconductor Industry Association. San Jose, CA, United States [Online] http://public.itrs.net/
14. Oshima, Y. "Legal Protection for Semiconductor Intellectual Property (IP)." *Asia and South Pacific Design Automation Conf. (ASP-DAC)* (January 2003) pp. 551–555.

15. Delp, G. S. "Reusing IP Increases the Leverage of Your Product Development Dollars. Chip Design Trends Reports." [Online] http://www.chipdesignmag.com/display.php?articleId=401
16. Maliniak, D. "EDA: Ticket to Super SoCs is Abstractions, IP Reuse." Electronic Design Online. April 29, 2002. [Online] http://electronicdesign.com/Articles/Index.cfm?AD=1&ArticleID=2136
17. Chandra, R. "IP-Reuse and Platform Base Designs." Design and Reuse Industry Articles. [Online] http://www.us.design-reuse.com/articles/6125/ip-reuse-and-platform-base-designs.html
18. Abrishami, R. "Vendors Must Support IP Reuse in SoC." *EE Times.* December 19, 2002. [Online] http://www.eetimes.com/story/OEG20021219S0018
19. Weekley, J. "Modeling Total Cost of Ownership for Semiconductor IP." Design and Reuse Industry Articles (2004). [Online] http://www.us.design-reuse.com/articles/9065/modeling-total-cost-of-ownership-for-semiconductor-ip.html
20. VSI Alliance. "The VSIA and IP Reuse." SOCcentral. August 2, 2005. [Online] http://www.soccentral.com/results.asp?EntryID=15373
21. VSI Alliance. [Online] http://vsia.org/index.htm
22. Scopus. "Scopus Overview: What is it?" [Online] http://www.info.scopus.com/about/
23. Canadian Intellectual Property Office (CIPO). "A Guide to Patents." (July 2004). [Online] http://strategis.ic.gc.ca/sc_mrksv/cipo/patents/patguide-e.pdf
24. United States Patent and Trademark Office (USPTO). "General Information Concerning Patents: Attorneys and Agents." (January 2005). [Online] http://uspto.gov/go/pac/doc/general/

Chapter 3

1. News Release. "Jack Kilby, Inventor of the Integrated Circuit, Dies at 81." Texas Instrument. [Online] http://focus.ti.com/pr/docs/preldetail.tsp?sectionId=594&prelId=c05034
2. Meindl, J. D., P. B. Meyers, G. Abraham, J. E. Iwersen, I. A. Lesk, and J. R. Roeder. "Definitions of Terms for Integrated Electronics." IEEE Standard, no. 274, effective December 1966, reaffirmed March 1980.
3. "The Chip that Jack Built." Texas Instrument. [Online] http://www.ti.com/corp/docs/kilbyctr/jackbuilt.shtml

4. Kang, S. M. and Y. Leblebici. *CMOS Digital Integrated Circuits—Analysis and Design.* New York: McGraw-Hill (2005).
5. Beeson, R. and H. Ruegg. "New Forms of all Transistor Logic." *ISSCC Digest of Tech. Papers,* (February 1962) pp. 10–11.
6. Rabaey, J. M., A. Chandrakasan, and B. Nikolic. *Digital Integrated Circuits—A Design Perspective.* Upper Saddle River, NJ: Pearson Education (2003).
7. Alvarez, A. R. *BiCMOS Technology and Applications.* Boston, MA: Kluwer Academic Publishers (1989).
8. Yeo, K. S., S. S. Rofail, and W. L. Goh, *CMOS/BiCMOS ULSI: Low-Voltage Low-Power.* Upper Saddle River, NJ: Prentice Hall (2002).
9. Ning, T. H. "Why BiCMOS and SOI BiCMOS?" IBM *J. Research and Development*, vol. 46, no. 2/3, (March/May 2002) pp. 181–186.
10. Davari, B., W. H. Chang, M. R. Wordeman, C. S. Oh, Y. Taur, K. E. Petrillo, D. Moy, J. J. Bucchignano, H. Y. Ng, M. G. Rosenfield, F. J. Hohn, and M. D. Rodriguez. "A High Performance 0.25 μm CMOS Technology," *International Electron Devices Meeting (IEDM) Tech. Digest* (1988) pp. 56–59.
11. Chang, W. H., B. Davari, M. R. Wordeman, Y. Taur, C. C.-H. Hsu, and M. D. Rodriguez. "High-Performance 0.25 CMOS Technology: I—Device and Characterization." *IEEE Trans. Electron Devices,* vol. 39, no. 4 (April 1992) pp. 959–966.
12. Sunter, S. "Analog, Digital, and Mixed-Signal People." *IEEE Design and Test of Computers,* vol. 16, no. 3 (July-September 1999) pp. 128.
13. Encyclopedia. "Analog." *EE Times.* [Online] http://www.eetimes.com/encyclopedia/defineterm.jhtml?term=analog
14. Misra, D. K. *Radio-Frequency and Microwave Communication Circuits—Analysis and Design.* New York: John Wiley & Sons, Inc. (2001).
15. Ellinger, F. *Radio Frequency Integrated Circuits and Technologies.* Leipzig, Germany: Springer (2007).
16. Mano, M. D. and M. D. Ciletti. *Digital Design.* Upper Saddle River, NJ: Pearson Education (2007).
17. Martin, G. and H. Chang. "System-on-Chip Design." *Proc. Int. Conf. ASIC* (October 2001) pp. 12–17.
18. Stevaert, M., A. H. M. Roermund, and J. H. Huijsing. *Analog Circuit Design: RF Circuits: Wide band, Front-Ends, DAC's, Design Methodology and Verification for RF and Mixed-Signal Systems, Low Power and Low Voltage.* The Netherlands: Springer (2006).

19. Handkiewicz, A. *Mixed-Signal Systems: A Guide to CMOS Circuit Design.* United States: Wiley-IEEE Press (2002).
20. Kester, W. *Mixed-signal and DSP Design Techniques.* Oxford, England: Newnes, Elsevier Science and Technology (2003).
21. Rizzoni, G. *Principles and Applications of Electrical Engineering.* New York: McGraw-Hill Higher Education (2004).
22. Hodges, D. A., H. G. Jackson, and R. A. Saleh. *Analysis and Design of Digital Integrated Circuits in Deep Submicron Technology.* New York: McGraw-Hill Higher Education (2004).
23. Intel. "Intel Unveils 16 Next-Generation Processors, Including First Notebook Chips Built on 45 nm Technology." Intel News Release. January 7, 2008. [Online] http://www.intel.com/pressroom/archive/releases/2008/20080107comp.htm
24. Rusu, S. "Trends and Challenges in VLSI Technology Scaling Towards 100nm." *European Solid-States Circuits Conf.* (September 2001) pp. 194–196.
25. Fildes, J. "Chips Pass Two Billion Milestone." BBC News. February 4, 2008. [Online] http://news.bbc.co.uk/1/hi/technology/7223145.stm
26. Castro-Lopez, R., F. V. Fernandez, O. Guerra-Vinuesa, and Rodriguez-Vazquez. *Reuse-Based Methodologies and Tools in the Design of Analog and Mixed-Signal Integrated Circuits.* The Netherlands: Springer (2006).
27. McConaghy, T. and G. Gielen. "Automation in Mixed-Signal Design: Challenges and Solutions in the Wake of the Nano Era." *IEEE/ACM Int. Conf. Computer-Aided Design* (November 2006) pp. 461–463.
28. Arsintescu, B. "Device Level Layout Optimization in Electronic Design Automation." U.S. Patent 6671866, December 30, 2003.

Chapter 4

1. Dholakia, J. "Reviewing Business Method Patents (BMP's): A Strategic Asset for Companies and Inventors." *Int. Business and Economics Research J.*, vol. 6, no. 1 (2007) pp. 49–62.
2. Chua, S. K. "Patenting Business Methods." *Law Gazette—An Official Publication of the Law Society of Singapore* (October 2001).
3. Guntersdorfer, M. "Software Patent Law: United States and Europe Compared." *Duke Law and Technology Review,* rev. 0006 (2003).

4. IPR Helpdesk. "Grace Period and Invention Law in Europe and Selected States." (February 2006). [Online] http://www.ipr-helpdesk.org/documents/GracePeriodinventionLaw_0000004514_00.xmlhtml/
5. Intellectual Property Office of Singapore. "About Patents." [Online] http://www.ipos.gov.sg/leftNav/pat/
6. International Bureau. "Industrial Applicability and Utility Requirements: Commonalities and Differences. (May 2003). [Online] http://www.wipo.int/edocs/mdocs/scp/en/scp_9/scp_9_5.pdf
7. Tysver, D. A. "Works Unprotected by Copyright Law." BitLaw—A Resource on Technology Law. [Online] http://www.bitlaw.com/copyright/unprotected.html
8. Reichman, J. H. "Legal Hybrids Between the Patent and Copyright Paradigms." *Columbia Law Review,* vol. 94, no. 8 (December 1994) pp. 2432–2558.
9. http://www.dictionary.reference.com/
10. World Intellectual Property Organization (WIPO). "Treaty on Intellectual Property in Respect of Integrated Circuits." Washington, DC. May 26, 2989. [Online] http://www.wipo.int/treaties/en/ip/washington/trtdocs_wo011.html
11. Agreement on Trade-Related Aspects of Intellectual Property Rights. April 15, 1994, Marrakesh Agreement Establishing the World Trade Organization, Annex 1C, Legal Instruments-Results of the Uruguay Round vol. 31, 33 I.L.M. 81 (1994).
12. WIPO Introductory Seminar on Intellectual Property. "General Introduction to Intellectual Property Rights." World Intellectual Property Organization (WIPO). April 19, 2004. [Online] http://www.wipo.int/edocs/mdocs/arab/en/wipo_ip_mct_apr_04/wipo_ip_mct_apr_04_2.pdf
13. Goldberg, M. D. "Global Dimensions of Intellectual Property Rights in Science and Technology." Book chapter: "Semiconductor Chip Protection as a Case Study." 329, 333 (Mitchel B. Wallerstein, et al., eds., 1993). Read online http://books.nap.edu/openbook.php?record_id=2054&page=329
14. Christie, A. *Integrated Circuits and Their Contents: International Protection.* London, United Kingdom: Sweet & Maxwell (1995).
15. Shih, T. "The Semiconductor Chip Protection Act of 1984: Is Copyright Protection for Utilitarian Articles Desirable?" *Computer/Law J.*, vol. 8, issue 2 (1986) pp. 129–201.
16. Samuelson, P. and S. Scotchmer. "The Law and Economics of Reverse Engineering." *The Yale Law J.*, vol. 111, no. 7 (May 2002) pp. 1575–1663.

17. Malisuwan, S. and J. Xu. "Industrial Property Rights Protection: Analysis of Integrated Circuits Act of Thailand." *Int. J. the Computer, the Internet and Management*, vol. 14, no. 3 (2006) pp. 1–7.
18. Samuelson, P. "Creating a New Kind of Intellectual Property: Applying Lessons of the Chip Law to Computer Programs." *Minnesota Law Rev.*, vol. 70: 471 (1985) pp. 471–531.
19. U.S. Department of Commerce, Patent and Trademark Office. "A Guide to Filing a Design Patent Application." [Online] http://www.uspto.gov/web/offices/pac/design/index.html#types
20. Renk, C. J. "Protect the Rights of Patent Owners." Forbes Commentary. March 31, 2008. [Online] http://www.forbes.com/2008/03/28/design-patent-rights-protection-oped-cx_cr_0331renk.html
21. World Intellectual Property Organization (WIPO). "About Industrial Designs." [Online] http://www.wipo.int/designs/en/about_id.html
22. BitLaw—A Resource on Technology Law. "Design Patents." [Online] http://www.bitlaw.com/patent/design.html
23. Bellis, M. "Design Patents vs. Other Types of Intellectual Property, Definition of Design." About.com. [Online] http://inventors.about.com/od/designpatents/a/design_patent.htm
24. Intellectual Property Office of Singapore. "About Designs." [Online] http://www.ipos.gov.sg/leftNav/des/
25. World Intellectual Property Organization (WIPO). "Understanding Industrial Property." [Online] http://www.wipo.int/freepublications/en/intproperty/895/wipo_pub_895.pdf
26. International Trademark Association (INTA). "Nontraditional Marks." [Online] http://www.inta.org/index.php?option=com_content&task=view&id=178&Itemid=59&getcontent=1
27. Moodie, A. M. "A Lion's Roar is Heard in Trademark Law." Fairfax Business Media. February 18, 2008. [Online] http://computerworld.co.nz/cio.nsf/84d556af64b8966fcc2572ad0046770e69617692db621f56cc2573f10030569d?OpenDocument
28. Intellectual Property Office of Singapore. "About Trademarks." [Online] http://www.ipos.gov.sg/leftNav/tra/
29. International Trademark Association (INTA). "What Do the Symbols ®, ™, and SM Mean?" International Trademark Association (INTA). [Online] http://www.inta.org/index.php?option=com_simplefaq&task=display&Itemid=0&catid=284&page=1&getcontent=5#FAQ53
30. Bellis, M. "What is the Proper Use of a Trademark Symbol?" About.com. [Online] http://inventors.about.com/od/inventing101trademarks/f/tm_symbol.htm.

31. Jordan, A. (23 "Radhard-By-Design Integrated Circuit Suppliers: A Success Story." Military Embedded Systems–Space Application (Editorial). March 23, 2006. [Online] http://www.mil-embedded.com/articles/authors/jordan/
32. D'Aliesio, G. "Reconfigurable Computing Platforms: Adapting for Space Applications. Military Embedded Systems–Space Application." (Editorial). March 2007. [Online] http://www.mil-embedded.com/pdfs/MacDonald.Mar07.pdf
33. Canadian Intellectual Property Office. "A Guide to Integrated Circuit Topographies." [Online] http://www.cipo.ic.gc.ca/epic/site/cipointernet-internetopic.nsf/vwapj/ictguide-e.pdf/$FILE/ictguide-e.pdf
34. Bereskin & Parr Intellectual Property Law. "About IP-Integrated Circuits." [Online] http://www.bereskinparr.com/English/about_ip/integrated_circuits_/aboutip_ic_1.html
35. Greguras, F. M. "Systems-on-a-Chip Intellectual Property Protection and Licensing Issues." Fenwick & West. May 16, 1998. [Online] http://www.fenwick.com/docstore/Publications/IP/IP_Articles/Systems-on-a-Chip.pdf
36. Matheson, T. G., J. F. Cicchiello, and G. Wikle. "Integrated Circuit Design Apparatus with Extensible Circuit Elements." U.S. Patent 5592392. January 7, 1997.
37. Yalcin, H., R. J. Palmero, K. A. Sakallah, M. S. Mortazavi, and C. Bamji. "Functional Timing Analysis for Characterization of Virtual Component Blocks." U.S. Patent 6457159. September 24, 2002.
38. Shuc, W. K. H. "Die Pad Crack Absorption System and Method for Integrated Circuit Chip Fabrication." U.S. Patent 6503820. January 7, 2003.
39. Rockman, H. B. *Intellectual Property Law for Engineers and Scientists.* New Jersey: John Wiley & Sons, Inc. (2004).
40. World Intellectual Property Organization (WIPO). "Is Computer Software Protected by Copyright?" [Online] http://www.wipo.int/copyright/en/faq/faqs.htm#P39_5114
41. Gottschalk vs. Benson, 409 U.S. 62, 72 (1972). [Online] http://caselaw.lp.findlaw.com/scripts/getcase.pl?court=US&vol=409&invol=63
42. Diamond vs. Diehr, 450 U.S. 175, 187 (1981). [Online] http://caselaw.lp.findlaw.com/scripts/getcase.pl?court=US&vol=450&invol=175
43. State St. Bank & Trust Co. vs. Signature Fin. Group, 149 F.3d 1368, 1373–77 (U.S. Fed. Cir. 1998).

44. WMS Gaming, Inc. vs. Int'l Game Tech., 184 F.3d 1339, 1348–49 (U.S. Fed. Cir. 1999) (quoting In re Alappat, 33 F.3d 1526, 1545 [U.S. Fed. Cir. 1994]).
45. Pratt, W. K., J. F. Abramatic, and O. Faugeras. "Method and Apparatus for Improved Digital Image Processing." European Patent 0005954, December 12, 1979.
46. European Patent Convention (EPC) 52(1) (1998).
47. European Patent Convention (EPC) 54(1).
48. BitLaw–A Resource on Technology Law. "Why Protect Software Through Patents." [Online] http://www.bitlaw.com/software-patent/why-patent.html
49. Maier, G. J. "Software Protection-Integrating Patent, Copyright and Trade Secret Law." *J. Patent and Trademark Office Society,* vol. 69, no. 3 (1987): 152–165.
50. Lai, J. Z., Y. B. Yu, L. Y. Dai, and M. L. Niu. "System and Method for Upgrading Firmware in a Display." U.S. Patent 20070002347, April 1, 2007.
51. Tech Interviews. "What is the Difference Between Firmware and Software?" [Online] http://qa.techinterviews.com/q/20060814153355AAE0NMh
52. International Bureau. "Standard St.9 Recommendation Concerning Bibliographic Data On and Relating to Patents and SPCS." WIPO Handbook on Industrial Property Information and Documentation. [Online] http://www.wipo.int/standards/en/pdf/03-09-01.pdf
53. United States Patent and Trademark Office (USPTO). "*Kind Codes* Included on USPTO Patent Documents." [Online] http://www.uspto.gov/web/forms/kindcodesum.html
54. Rockman, H. B. *Intellectual Property Law for Engineers and Scientists.* New Jersey: John Wiley & Sons, Inc. (2004).
55. Gordon, T. T. and A. S. Cookfair. "Patent Fundamentals for Scientists and Engineers." United States: Lewis Publishers, CRC Press (2000).
56. United States Patent and Trademark Office (USPTO). "Title and abstract: Title of invention. Title 37 of Code of Federal Regulations," Section 1.72(a). [Online] http://www.uspto.gov/web/offices/pac/mpep/documents/0600_606.htm
57. Eisenberg, H. M. "Patent law you can use™: Reading a Patent, Part I The Cover Page." Yale University Office of Cooperative Research (2000). [Online] http://www.yale.edu/ocr/pfg/guidelines/patent/reading_patent1_cover.html

58. World Intellectual Property Organization (WIPO). "International Patent Classification" (2006). [Online] http://www.wipo.int/classifications/ipc/ipc8/guide/en/guide.pdf
59. United States Patent and Trademark Office (USPTO). "Title and Abstract: Abstract of the Disclosure. Title 37 of Code of Federal Regulations," Section 1.72(b). [Online] http://www.uspto.gov/web/offices/pac/mpep/documents/0600_608_01_b.htm
60. http://patft.uspto.gov/
61. http://www.uspto.gov/patft/index.html
62. http://www.wipo.int/pctdb/en/
63. http://www.delphion.com/
64. http://www.getthepatent.com/
65. United States Patent and Trademark Office (USPTO). "Guide for Preparation of Patent Drawings." (June 2002). [Online] http://www.uspto.gov/web/navaids/siteindx.htm
66. Stim, R. *Patent, Copyright, and Trademark: An Intellectual Property Desk Reference.* Berkeley, CA: Nolo (2006).
67. United States Patent and Trademark Office (USPTO). "Summary of the Invention. Title 37 of Code of Federal Regulations," Section 1.73. [Online] http://www.uspto.gov/web/offices/pac/mpep/documents/0600_608_01_d.htm#sect608.01d
68. Carr, F. K. *Patents Handbook: A Guide for Inventors and Researchers to Searching Patent Documents and Preparing and Making an Application.* United States: McFarland & Company (1995).
69. United States Patent and Trademark Office (USPTO). "Specification-Patent Laws. Title 35 of the United States Code," Section 112 [Online] http://uspto.gov/web/offices/pac/mpep/documents/appxl_35_U_S_C_112.htm

Chapter 5

1. Some countries might call it "topographies" (e.g., Canada's Integrated Circuit Topography Act, 1990) or *mask work* (e.g. as defined in the United States' Title 17 USC on Copyrights, chapter 9 of which covers "Protection for Semiconductor Chip Products").
2. Sourced from the Intellectual Property Office of Singapore website. http://www.ipos.gov.sg/leftNav/aboip/
3. Sections 2(1) and 19(2)(a) of the Patents Act.

4. Section 46(1) of the Patents Act.
5. Section 46(1) of the Patents Act.
6. Section 19(2) of the Patents Act.
7. Section 36 of the Patents Act.
8. Section 36A of the Patents Act.
9. Section 36A(4) of the Patents Act.
10. The United States has a Copyright Office that allows voluntary registration of copyright.
11. Section 26(1) of the Copyright Act.
12. Section 14 of the Copyright Act.
13. Section 27(2) of the Copyright Act.
14. Section 24(5) of the Copyright Act.
15. Section 7(1) of the Copyright Act.
16. Section 7(1) of the Copyright Act.
17. Section 7(1) of the Copyright Act.
18. Section 7A of the Copyright Act.
19. Section 7(1) of the Copyright Act.
20. Section 30(6) of the Copyright Act.
21. Section 30(4) of the Copyright Act.
22. Section 30(5) of the Copyright Act.
23. Section 30(3) of the Copyright Act.
24. Section 7(1) of the Copyright Act.
25. Section 28(2) of the Copyright Act.
26. Section 28(3) of the Copyright Act.
27. Section 7(1) of the Copyright Act.
28. Section 92 of the Copyright Act.
29. Section 7(1) of the Copyright Act.
30. Section 93 of the Copyright Act.
31. Section 7(1) of the Copyright Act.
32. Section 94 of the Copyright Act.
33. Section 7(1) of the Copyright Act.
34. Section 95 of the Copyright Act.
35. Section 96 of the Copyright Act.
36. Section 6(1) of the Layout-Designs of Integrated Circuits Act.

37. Section 6(1) of the Layout-Designs of Integrated Circuits Act.
38. Section 8 of the Layout-Designs of Integrated Circuits Act.
39. Section 5 of the Layout-Designs of Integrated Circuits Act.
40. This paragraph b(ii) offers greater clarity on the *combination* qualification compared to certain countries' equivalent provisions, such as section 5(2)(b) of Malaysia's Layout-Designs of Integrated Circuits Act (LODintegrated circuitA) 2000 and section 5(1)(b) of Singapore's LOD integrated circuit A, rev. 2000 in which the repeated use of the words *combination* and *commonplace* before and after the criterion being defined only serves to confuse. By comparison, Article 2(2) of the EEC Council Directive No. 87/54 is relatively clearer in stating, "Where the topography of a semiconductor product consists of elements that are commonplace in the semiconductor industry, it shall be protected only to the extent that the combination of such elements, taken as a whole, fulfils the abovementioned conditions *of intellectual effort and noncommonplace.*"
41. For example, Singapore prescribes on February 15, 1999, the date of coming into force of its Layout-Designs of Integrated Circuits Act. Malaysia does not prescribe any such commencement of protection for independently created identical layout-designs.
42. This independent creation requirement is deemed redundant as it would be an inherent quality of originality or own intellectual effort. For this reason, presumably, it is not listed as a separate requirement in 37 USC 902(b) and in the EEC Council Directive No. 87/54. As for identical layout-designs, it is also to be understood that when there are many layout-designs that are identical, it is thus established that it has become commonplace for that particular layout-design.
43. Section 5(4) specifically requires reduction to a material form which includes *recorded in documentary form.*
44. Section 2 of the Layout-Designs of Integrated Circuits Act.
45. Section 6(1) of the Layout-Designs of Integrated Circuits Act.
46. Section 6(1) of the Layout-Designs of Integrated Circuits Act.
47. Section 6(1) of the Layout-Designs of Integrated Circuits Act.
48. Section 7 of the Layout-Designs of Integrated Circuits Act.
49. Section 7 of the Layout-Designs of Integrated Circuits Act.
50. Sections 15(2) and 18(1) of the Trade Marks Act.

Chapter 6

1. See for instance *Institut Pasteur & Anor v. Genelabs Diagnostics Pte Ltd & Anor [2000] SGHC 53.*
2. See for instance *Trek Technology (Singapore) Pte Ltd v. FE Global Electronics Pte Ltd and Others and Other Suits (No 2) [2005] 3 SLR 389.*
3. Ibid p. 172.
4. Loon, N. L. W. *Law of Intellectual Property of Singapore* (2008): 411.
5. Ibid.
6. *Terrell on the Law of Patents* (2006): 326.
7. Ibid.
8. Ibid.
9. Ibid at pg. 327.
10. Loon, N. L. W. *Law of Intellectual Property of Singapore* (2008): 412.
11. Ibid at pg. 412–413.
12. Ibid at pg. 419.
13. Ibid at pg. 411.
14. Ibid at pg. 416.
15. Ibid.
16. Section 67 of the Patents Act states that "civil proceedings may be brought in the court by the proprietor of a patent in respect of any act alleged to infringe the patent."
17. Section 74(1) of the Patents Act.
18. Section 75 states: "Where by virtue of a transaction, instrument, or event to which section 43 applies a person becomes the proprietor or one of the proprietors or an exclusive licensee of a patent and the patent is subsequently infringed, the Court or the Registrar shall not award him damages or order that he be given an account of the profits in respect of such a subsequent infringement occurring before the transaction, instrument or event is registered unless:
    (a) the transaction, instrument, or event is registered within the period of 6 months beginning with its date; or
    (b) the Court or the Registrar is satisfied that it was not practicable to register the transaction, instrument, or event before the end of that period and that it was registered as soon as practicable thereafter."
19. Section 67(2) of the Patents Act.
20. Section 69(1) of the Patents Act.

21. Section 69(3)(a) of the Patents Act.
22. Section 69(3)(b) of the Patents Act.
23. Section 69(3)(c) of the Patents Act.
24. Section 69(4) of the Patents Act.
25. Section 2 of the Patents Act defines *Court* as the *High Court*.
26. Since September 2002.
27. Section 7A(1)(b) of the Copyright Act; *Aztech Systems Pte Ltd v. Creative Technology Ltd [1996] 1 SLR 683* at 700.
28. *Television Broadcasts Ltd v. Golden Line Video & Marketing Pte Ltd [1988] SLR 930*.
29. Ibid.
30. Section 31(1) of the Copyright Act.
31. Wei, G. *The Law of Copyright in Singapore.* SNP Editions (2000): 497.
32. Ibid.
33. Ng-Loy, W. L., *Law of Intellectual Property of Singapore.* Singapore: Sweet & Maxwell Asia (2008): 139.
34. Section 33(1) of the Copyright Act.
35. *Television Broadcasts Ltd v. Golden Line Video & Marketing Pte Ltd [1988] SLR 930* at 30.
36. Section 119(1) of the Copyright Act.
37. Section 119(2) of the Copyright Act.
38. Section 8 of the Layout-Designs of Integrated Circuits Act.
39. Section 12(2) of the Layout-Designs of Integrated Circuits Act.
40. Hodkinson, K. *Protecting and Exploiting New Technology and Designs.* Taylor & Francis (1987): 331.
41. Section 13(1) of the Layout-Designs of Integrated Circuits Act.
42. Section 14(1) of the Layout-Designs of Integrated Circuits Act.
43. Sections 13(2) and 14(2) of the Layout-Designs of Integrated Circuits Act.
44. Section 14(5) of the Layout-Designs of Integrated Circuits Act.
45. Section 11 of the Layout-Designs of Integrated Circuits Act.
46. Section 11(2) of the Layout-Designs of Integrated Circuits Act.
47. Matsura, J. H. *Managing Intellectual Assets in the Digital Age.* Artech House Inc. (2003): 26.
48. http://www.edb.gov.sg/edb/sg/en_uk/index/why_singapore/singapore_rankings.html

49. Section 26(1) of the Trade Marks Act.
50. Tan Tee Jim, S. C. *Law of Trade Marks and Passing Off in Singapore.* Sweet & Maxwell Asia (2005): 233.
51. Section 8 of the Trade Marks Act.
52. Section 31(2) of the Trade Marks Act.
53. See for instance, *Econlite Manufacturing Pte Ltd v Technochem Holdings Pte Ltd* (1994) 2 SLR 454 where the Court was of the view that as there was no evidence that the plaintiffs had suffered any substantial or serious loss or damage, the prospect of recovering damages of a significant amount was too slight to justify the ordering of an enquiry. Hence, there was no order for an enquiry as to damages.
54. Ibid 2 at pg 290.
55. Section 31(6) of the Trade Marks Act provides factors to be taken into account to quantify the statutory damages, such as: the flagrancy of the infringement of the registered trademark, any loss that the plaintiff has suffered or is likely to suffer by reason of the infringement, any benefit shown to have accrued to the defendant by reason of the infringement, the need to deter other similar instances of infringement, and other relevant matters.
56. Section 31(5) of the Trade Marks Act.
57. Section 32(1)(a) of the Trade Marks Act.
58. Section 32(1)(b) of the Trade Marks Act.
59. Section 32(3) of the Trade Marks Act.
60. Section 35(2) of the Trade Marks Act.
61. Section 35(1)(a)(b) of the Trade Marks Act.

Chapter 7

1. Section 25 of the Singapore Patents Act; Rules 19, 21-23 of the Singapore Patents Rules.
2. Section 26(8) of the Singapore Patents Act; Rule 26(5) of the Singapore Patents Rules.
3. Section 26(1) of the Singapore Patents Act.
4. Section 27 of the Singapore Patents Act; Rule 29(1) of the Singapore Patents Rules.
5. Section 2(1) of the Singapore Patents Act.
6. Rule 41 of the Singapore Patents Rules.
7. Section 29(2)(c)(ii) of the Singapore Patents Act.

8. Rule 46(4) of the Singapore Patents Rules.
9. Section 30 of the Singapore Patents Act; Rule 41 of the Singapore Patents Rules.
10. Section 34 of the Singapore Patents Act.
11. Section 34(1)(a) of the Singapore Patents Act.
12. Article 1 of the PCT.
13. Article 9(1) of the PCT; Rule 18.1 of the Regulations under PCT.
14. Article 10 of the PCT; Rule 19.1 of the Regulations under PCT.
15. Article 9 of the PCT; Rule 18.3 of the Regulations under PCT.
16. Article 11 of the PCT; Rule 20 of the Regulations under PCT.
17. Rules 15-16 of the Regulations under PCT.
18. Article 21(2)(a) of the PCT.
19. Article 18 of the PCT; Rules 42-43 of the Regulations under PCT.
20. Article 19 of the PCT.
21. Article 34 of the PCT.
22. Rule 66.1*bis.* (b) of the Regulations under PCT.
23. Article 36 of the PCT; Rule 70 of the Regulations under PCT.
24. Rule 44*bis.* 1 (a) of the Regulations under PCT.
25. Rule 73.2 of the Regulations under PCT.
26. Section 5(1)(2)(3) of the Singapore Trade Marks Act.
27. Section 5(1)(2)(3) of the Singapore Trade Marks Act.
28. Section 5(4) of the Singapore Trade Marks Act.
29. Section 8 of the Singapore Trade Marks Act.
30. Section 12(3) of the Singapore Trade Marks Act.
31. Section 12(3) of the Singapore Trade Marks Act.
32. Section 13 of the Singapore Trade Marks Act.
33. Section 19 of the Singapore Trade Marks Act.
34. Section 22 of the Singapore Trade Marks Act.
35. Article 1 of Madrid Protocol.
36. Article 2 of Madrid Protocol.
37. Rule 14(1) of Madrid Protocol Common Regulations.
38. Article 4(1)(a) of Madrid Protocol.
39. Section 34 of the Singapore Patents Act.

Index